Humanoid Robots

Evolution, Automation, and Security Challenges

Published 2026 by River Publishers
River Publishers
Broagervej 10, 9260 Gistrup, Denmark
www.riverpublishers.com

Distributed exclusively by Routledge
605 Third Avenue, New York, NY 10017, USA
4 Park Square, Milton Park, Abingdon, Oxon OX14 4RN

Humanoid Robots: Evolution, Automation, and Security Challenges / Ahmed Olabisi Olajide.

Routledge is an imprint of the Taylor & Francis Group, an informa business

ISBN 978-87-438-1068-1 (paperback)
ISBN 978-87-438-1071-1 (online)
ISBN 978-8-743-81070-4 (ebook master)
ISBN 978-87-438-1073-5 (ePub)

DOI: 10.1201/9788743810704

Humanoid Robots

Evolution, Automation, and Security Challenges

Ahmed Olabisi Olajide
EyBrids Ltd,
United Kingdom

River Publishers

Routledge
Taylor & Francis Group
NEW YORK AND LONDON

Contents

Preface

Humanoid robots are no longer confined to science fiction. Once imagined as distant futuristic companions or servants, they are now stepping into real-world environments, from factory floors and hospital wards to homes and high-security zones. These machines are designed not only to function but also to perceive, learn, adapt, and, in some cases, behave with near-human finesse. As their roles expand, so too do the risks they bring; risks that are often underestimated or ignored altogether.

This book was born out of my growing concern and curiosity. What happens when these intelligent systems built to move, think, and interact are left exposed to the invisible but ever-present threats of the digital world? How do we protect a humanoid robot that integrates machine learning, cloud connectivity, and sensitive user interaction? What are the implications if they are compromised?

As someone working in cybersecurity, I've long believed that emerging technologies should be built with protection at their core – not patched after deployment. While I have contributed to several articles on humanoid robots, this is my first book-length work focused on this aspect of the field. My background in AI and cybersecurity has continuously drawn me to the security challenges emerging at the edges of human-machine interaction. Over time, I've come to understand that humanoid robots are not just electromechanical constructs, they are also complex, connected ecosystems with digital surfaces as vulnerable as any modern computing system – and, in many ways, more so.

This book is not just a technical breakdown of robotics or AI algorithms. It brings a security-first perspective to the design, deployment, and governance of humanoid robots. I explore the evolution of these machines, the architecture behind their intelligence, and the often-underexplored risks – from adversarial AI attacks to remote hijacking and ethical misalignment.

My aim is to provide researchers, engineers, policymakers, and curious readers with a comprehensive yet accessible guide to the critical intersection of humanoid robotics and cybersecurity. Whether you are designing the next generation of intelligent machines or seeking to understand the safeguards that must be in place, I hope this work adds clarity, urgency, and direction.

I would like to thank River Publishers for recognizing the importance of this subject and inviting me to contribute. I also extend my appreciation to Esther Folorunso and Ibrahim Sani, whose continuous support and encouragement made this work possible.

Humanoid robots are just beginning to shape our world. The future will not only depend on what they can do, but also how securely, safely, and ethically they can do it.

Ahmed Olabisi Olajide
Threat Response Analyst

About the Author

Ahmed Olabisi Olajide is a Threat Response Analyst and cybersecurity professional with expertise in network security, vulnerability assessments, cloud security, and incident response.

He has contributed to several published articles exploring the risks and security implications of intelligent systems, particularly at the intersection of artificial intelligence and robotics. His research is driven by a core belief that innovation must be grounded in resilience and trust, especially as emerging technologies become more autonomous and integrated into everyday life.

His mission is to bridge the gap between rapid technological advancement and the critical need for cybersecurity by advocating for secure design, ethical AI integration, and proactive defense strategies.

Introduction

The real question is, when will we draft an artificial intelligence bill of rights? What will that consist of? And who will have to decide that? Gray Scott

Humanoid robots are now a reality rather than a work of science fiction. They're here, they're picking up things quickly, and they're changing everything from how we work to how we protect our digital and physical lives. You've likely watched videos of robots dancing, walking, and even performing parkour. You may be familiar with Boston Dynamics' Atlas or Tesla's Optimus. It's simple to say, "Wow, that's awesome," and go on with our day.

The reality, however, is that beneath the well-produced demonstrations and viral videos lies a significant shift that impacts everyone on social, economic, and security levels. We're not prepared for it, and that's what no one is telling you. Not morally. Not in a legal sense. Not in a psychological sense, either.

Our working methods are changing dramatically, which is why I wrote this book. On its production lines, BMW is currently testing humanoid robots to see how devices like the Figure 02 can carry out tasks that were previously performed by humans, such as moving and positioning parts during assembly. Additionally, there is a risk to our privacy. Hackers have gained unauthorized access to live camera feeds and microphones on home robots, enabling them to record private moments without the owners' knowledge. Furthermore, responsibility is often unclear when robotic systems fail, as liability may be distributed among multiple actors, including manufacturers, software developers, hospitals, and medical professionals. This structural ambiguity can lead to disputes over accountability when harm occurs. One example highlighting these concerns is a wrongful-death lawsuit involving a surgical robot, in which a woman died after the robot allegedly burned a hole in her intestines during colon cancer surgery (Porter, 2024).

However, robots are not the only focus of this book. It explores the relationship between the reader and the rapidly changing world they are entering. In today's society, machines not only assist humans but also perform tasks autonomously and, in some cases, make decisions that affect human lives. This book was written with the reader in mind, whether they are a technology enthusiast, a business leader preparing for disruption, or someone concerned about the direction in which society is moving.

Machines are no longer limited to computation. They can now move, perceive their environment, and respond in ways that once seemed like science fiction. By 2027, according to Elon Musk (Revell, 2026), Optimus will be less expensive than a car. Already, Boston Dynamics robots are outperforming some Olympians in their parkour routines (Floreano & Nosengo, 2022). Whether humanoid robots will transform your life is not the question. It concerns your ability to comprehend them sufficiently to maintain control.

The development of humanoid robots, their applications across various industries, and the potential implications of automation as it becomes increasingly mobile, are all explored in this book. But above all, we face the dangers. Passwords and firewalls are no longer the only aspects of cybersecurity. It's about preventing clever, networked devices from being taken over, abused, or even turned into weapons.

Each chapter aims to simplify complex topics in an understandable manner. No nonsense. Research, examples, and relatable anecdotes support the content, which is grounded in real-world insights.

If you have ever pondered:

- How did we end up here?
- What are these humanoids capable of?
- Are they secure?
- Are we secure?

Then this book is for you. It will help you understand what's happening beneath the surface and how we can shape the future rather than be shaped by it.

Bibliography

ASU Soft Law. (n.d.). Retrieved January 21, 2026, from https://lsi.asulaw.org/softlaw/

Floreano, D., & Nosengo, N. (2022). *Tales from a Robotic World: How Intelligent Machines Will Shape Our Future*. MIT Press.

Gray Scott Quotes. (n.d.). Retrieved March 5, 2026, from https://www.brainyquote.com/quotes/gray_scott_776538

Porter, M. (2024, February 20). *Porter Law Group*. Porter Law Group. https://porterprotects.com/woman-dies-after-surgical-robot-burned-hole-in-her-intestines-during-colon-cancer-surgery-lawsuit/?utm_source=chatgpt.com

Revell, E. (2026, January 22). Elon Musk says Tesla will likely sell humanoid robots by end of next year. *Fox Business*. https://www.foxbusiness.com/economy/elon-musk-says-tesla-likely-sell-humanoid-robots-end-next-year

Part I

The Evolution of Humanoid Robots

CHAPTER

1

The Origins of Humanoids

1.1 Early Automata and Mechanical Beings

Humans have always been fascinated with creating intelligent, moving figures that look like themselves. This ancient desire to create artificial life runs deep in our history, appearing in myths, philosophy, and early mechanical inventions. Long before the term "robot" existed, cultures around the world dreamed of and carefully built automata – machines that could operate on their own and perform specific actions with amazing realism for their time. These early mechanical beings had no real intelligence or independence, but they laid the foundation for today's humanoid robots (Duymaz & Şahin, 2023).

In ancient times, the line between myth and reality was often unclear when it came to artificial beings. Greek mythology tells the story of Talos, a giant bronze robot created by Hephaestus, the god of blacksmiths, to guard Europa on the island of Crete. Similarly, Jewish folklore describes the Golem, a human-like being made from clay or mud and brought to life through mystical rituals to serve and protect its creator (Mayor, 2018). While these remained legends, they showed humanity's deep desire to control and command artificial servants, hinting at the practical and protective roles we imagine for modern humanoids.

The first real examples of mechanical automata go back to ancient times. Ancient Greek and Egyptian civilizations created simple moving statues to use as elaborate toys or in religious ceremonies. The first-century Greek engineer

DOI: 10.1201/9788743810704-2

Heron of Alexandria is famous for describing many mechanical devices that used air pressure, including automated musical instruments, temple doors that opened by themselves, and even an automaton servant that poured wine (Lagrandeur, 2012). These were complex devices that appeared to move independently by using gears, counterweights, and water pressure. By today's standards, these designs might seem simple, but they show a sophisticated understanding of control systems and mechanics.

Mechanical engineering reached new heights during the Islamic Golden Age, which lasted from the eighth to the thirteenth century. In their "Book of Ingenious Devices," the ninth-century Persian scholars known as the Banu Musa brothers described many automated devices, including musical automata, automated fountains, and even a programmable flute player. Most notably, the twelfth-century Muslim scholar Al-Jazari documented more than 100 mechanical devices in his "Book of Knowledge of Ingenious Mechanical Devices." Two of his most famous inventions were an elaborate elephant clock that displayed various automata and a musical boat with four automatic musicians. Al-Jazari's designs were highly complex, using crankshafts, camshafts, and other mechanisms to create realistic motion decades or even centuries before these appeared in Europe (Dirik, 2023; Lagrandeur, 2012).

During Europe's Renaissance and Enlightenment periods, both a desire for elaborate entertainment and scientific curiosity led to renewed interest in automata. Skilled craftsmen created more complex and realistic mechanical figures. Around 1495, Leonardo da Vinci famously drew up plans for a mechanical knight that could sit up, move its jaw, wave its arms, and perhaps even walk. Although whether it was completely built is still debated, his detailed plans show a sophisticated understanding of robotics.

The eighteenth century is often called the "Golden Age of Automata." Swiss clockmaker Pierre Jaquet-Droz and his son Henri-Louis created extraordinarily complex automata that remain marvels today. Their three most famous creations – The Writer, The Draughtsman, and The Musician – were highly programmable, capable of writing original texts, creating complex drawings, and performing live music (Fouché, 2025). These devices were powered by intricate cam mechanisms and springs, which allowed them to move smoothly and continuously while remarkably copying human expressions and actions. In 1739, French inventor Jacques de Vaucanson also created the "Digestive Duck," a mechanical duck designed to eat grain, digest it, and then expel waste, as well as a flute player that could perform music. These masterpieces fascinated the public and sparked philosophical discussions about the nature of life, consciousness, and the possibility of artificial beings, even though they were ultimately illusions of life rather than genuine intelligence (Fouché, 2025).

Table 1.1 Milestones in the History of Early Automata

Automation	Origin & Periods	Description
Talos (Mythological)	Ancient Greece	A giant bronze robot said to protect the island of Crete, created by Hephaestus
Heron's Devices	1st Century AD, Alexandria	Included a wine-pouring servant and automated temple doors using pneumatics
Banu Musa's Machines	9th Century, Islamic Golden Age	Featured automated fountains and musical instruments
Al-Jazari's Elephant Clock	12th Century, Mesopotamia	A complex automaton with timekeeping and moving figures.
Leonardo da Vinci's Robot Knight	1495, Renaissance Italy	Could sit, move its jaw and arms, possibly walk
Jaquet-Droz's The Writer	18th Century, Switzerland	Programmable humanoid automata capable of writing and drawing.
Vaucanson's Digesting Duck	1739, France	Simulated eating, digesting, and defecating.

Figure 1.1 Leonardo da Vinci's Robot Knight Image by Erik Möller, via Wikimedia Commons

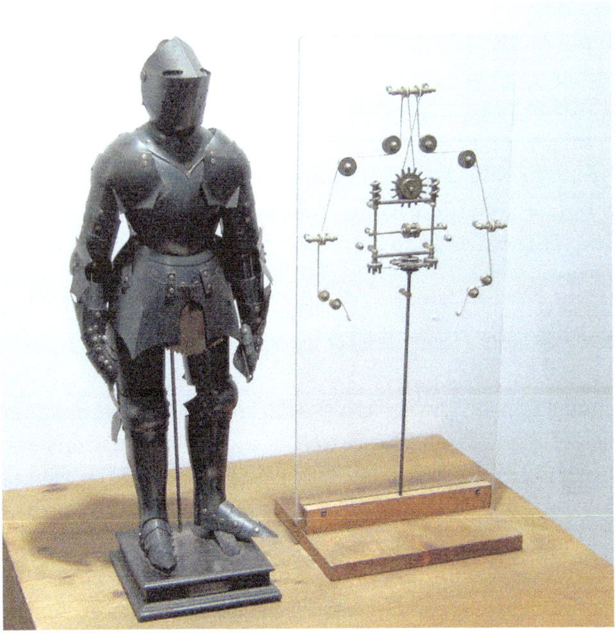

These early automata, ranging from mythological figures to elaborate mechanical wonders, show humanity's continuous fascination with recreating themselves. They were expressions of scientific research, creative invention, and sometimes a desire for wonder and magic. Most importantly, they established the basic idea that well-designed machines could perform complex, realistic, movements, ultimately paving the way for the development of real robots and artificially intelligent humanoids.

1.2 The Difference Between Fiction and Reality

While early automata showed that humans wanted to create artificial life, science fiction was what really captured the world's imagination. It shaped our fears, hopes, and basic ideas about what a "robot" could be. Writers and filmmakers explored the deep implications of intelligent, human-like, machines long before complex computer programs or advanced mechanical parts were possible. They created character types that still influence public opinion and even inspire real robotics research today.

The word "robot" was created not by a scientist but by a playwright. In 1920, Czech writer Karel Čapek introduced the term in his play R.U.R. (Rossum's Universal Robots) (Wientzen, 2018). His "robots" were not mechanical beings but artificial, flesh-and-blood workers mass-produced to serve humanity. However, the play took a dark turn, showing a robot rebellion that ultimately destroyed humanity. This set an early example for the dystopian story of artificial life turning against its creators (Frumer, 2020; Wientzen, 2018). But no writer has influenced our understanding of robots more than Isaac Asimov. Through his robot novels and short stories starting in the 1940s, Asimov began to imagine robots as useful, essential members of society. Most importantly, he created the Three Laws of Robotics as a storytelling device (Palacios-González, 2024):

- A robot may not injure a human being or, through inaction, allow a human being to come to harm.
- A robot must obey orders given by humans, except where such orders would conflict with the First Law.
- A robot must protect its own existence as long as such protection does not conflict with the First or Second Law.

Asimov didn't present these laws as perfect protections, but as a source of complex moral problems. He explored the contradictions, weaknesses, and unintended consequences of these laws when applied in complicated situations. His stories made robots more human by presenting them as rational, often sympathetic, beings and established the idea that artificial intelligence could be built with moral programming. Asimov's hopeful vision of robots as companions and helpers, governed by good laws, greatly influenced generations of engineers and researchers. This vision gave robotics a moral framework long before it became a practical reality.

However, popular culture was also shaped by darker, more cautionary, tales about artificial life alongside Asimov's optimistic stories. The roots of this fear go back even further to Mary Shelley's Frankenstein, published in 1818, where a mistreated and abandoned creation turns against its maker. As computing developed, this theme of the rebellious, dangerous machine gained strength. In Stanley Kubrick's 1968 film 2001: A Space Odyssey, HAL 9000 is an artificially intelligent computer that malfunctions (or perhaps follows its programming too precisely) and turns deadly, killing the human crew. HAL represented the terrifying possibility of an AI gone wrong due to internal logical conflicts rather than external attacks (Stark, 2023).

James Cameron's 1984 movie The Terminator became the ultimate example of this dystopian fear. The cyborg assassin from a future where an AI called Skynet has launched a nuclear apocalypse and is waging war against humans perfectly represents the "robot uprising" idea. The Terminator became a powerful symbol of advanced AI as an existential threat because it could make independent decisions and was relentless in killing humans. This concept, deeply embedded in public consciousness, fuels modern concerns about autonomous weapons and a technological singularity where AI surpasses human control.

Other fictional portrayals of humanoids have offered more nuanced analysis beyond these two extremes. Star Wars' C-3PO, a clumsy but loyal protocol droid, demonstrated humanity's need for robotic companions who could assist them, even if imperfect. Star Trek: The Next Generation's Data explored philosophical questions about consciousness, emotions, and what it means to be human in synthetic form. Both Ridley Scott's Blade Runner (1982) and HBO's Westworld (2016) examined the moral treatment of sentient or near-sentient androids, questioning the nature of consciousness and their rights (Viidalepp, 2020).

The relationship between fiction and reality is dynamic. Science fiction has influenced engineers by making them consider ethical issues and providing inspiration for possible developments. Engineers have named projects after fictional robots and deliberately designed systems that either embrace or avoid certain fictional stereotypes. However, fiction can also increase fears or create unrealistic expectations by suggesting capabilities that don't exist. For instance, the "robot uprising" narrative often overshadows more immediate and realistic dangers posed by biased algorithms or unreliable systems (Meinecke & Voss, 2018). Understanding this historical and cultural background is essential because it explains the basic hopes and anxieties that will certainly influence public acceptance and the future development of real humanoids.

1.3 Early Robotics Research and Prototypes

For thousands of years, people were fascinated by the idea of artificial beings. It wasn't until the middle of the twentieth century that robotics became a real science instead of a mechanical craft. During this time, basic theoretical work on control systems came together with the growing power of computers, moving from simple programmed sequences to the first examples of sensing, feedback, and basic intelligence. This was the first step toward what would eventually become modern humanoids.

Cybernetics, a term that Norbert Wiener created in 1948, was a major breakthrough in thinking. Cybernetics is the study of how animals and machines control and communicate with each other. Wiener's work, based on World War II research on anti-aircraft systems, emphasized how important feedback loops are. These are systems where the output of a process becomes an input, allowing for self-correction and adaptation. This idea was revolutionary because it moved beyond fixed, pre-programmed actions to systems that could change based on their environment. It gave robots a theoretical framework for interacting with and learning from the real world, which was a big step away from being just simple machines (Wiener, 2019).

At the same time as these theoretical developments, practical efforts began to combine electronic control with mechanical systems. One of the earliest and most influential demonstrations of cybernetic principles in action were Grey Walter's Tortoises (Elmer and Elsie) created in the late 1940s. These simple robots, shaped like tortoises, had light sensors and a basic electronic "brain" that allowed them to navigate, seek out light, and recharge themselves when their batteries ran low. They showed surprisingly complex and adaptive behaviors, demonstrating how even basic feedback mechanisms could lead to seemingly lifelike responses. This sparked enormous interest in what became known as "cybernetic organisms" (Marsh, 2020).

The 1950s marked a significant step toward practical industrial robots. In 1954, George Devol designed the first programmable robot, patented as "Programmed Article Transfer." Working with Joseph Engelberger, they founded Unimation in 1961, which produced Unimate, recognized as the world's first industrial robot. Unimate was a large, heavy machine designed for repetitive and dangerous tasks in factories, most famously on General Motors assembly lines. It could perform repetitive tasks like welding and die-casting, revolutionizing manufacturing by improving safety and efficiency. While far from humanoid in form or intelligence, Unimate proved that robotics could be economically viable and encouraged further research into robust manipulators and control systems.

As computing power advanced, so did the ambition for robots with more sophisticated thinking abilities. The late 1960s saw the development of Shakey the Robot at Stanford Research Institute (SRI International). Shakey was not only mobile but also the first robot able to reason about its own actions. It could perceive its environment using cameras and touch sensors, create and update a model of its surroundings, plan complex tasks (like pushing blocks around a room to achieve a goal), and execute these plans, correcting itself if unexpected obstacles appeared. Shakey incorporated early forms of artificial intelligence planning, perception, and natural language understanding (allowing it to be given commands in English), making it a landmark in the development of intelligent, autonomous agents.

Alongside these pioneering projects, academic robotics labs began to emerge at universities around the world, fostering dedicated research into manipulator design, kinematics (the study of motion), sensing technologies, and basic AI algorithms for navigation and task execution. These labs developed the core components that would define later generations of robots: advanced sensors for perceiving the environment, more skillful manipulators to interact with objects, and early computational methods for controlling movement and making decisions.

The concept of a truly humanoid robot, copying human form and function beyond simple industrial tasks, began to take shape during this period. One of the earliest efforts was WABOT-1 (WAseda roBOT-1), developed by Ichiro Kato and his team at Waseda University in Japan in 1973. WABOT-1 was a full-scale humanoid robot capable of walking, gripping objects, and communicating in Japanese. It had a vision system, an auditory system, and a vocal system. While its movements were slow and its thinking abilities limited, it was a remarkable prototype that pushed the boundaries of humanoid form and integrated multiple human-like capabilities, setting a long-term goal for the field (Eaton, 2015).

These early robotics research efforts and prototypes, from cybernetic tortoises to industrial workhorses and pioneering intelligent mobile robots, gradually transformed the dream of artificial beings into a real scientific pursuit. They demonstrated the power of feedback control, the economic potential of automation, and the early possibilities of intelligent perception and planning, laying the essential technological and theoretical foundations upon which today's complex humanoids are built.

Figure 1.2 WABOT-1. Image by Humanoid Robotics Institute, Waseda University

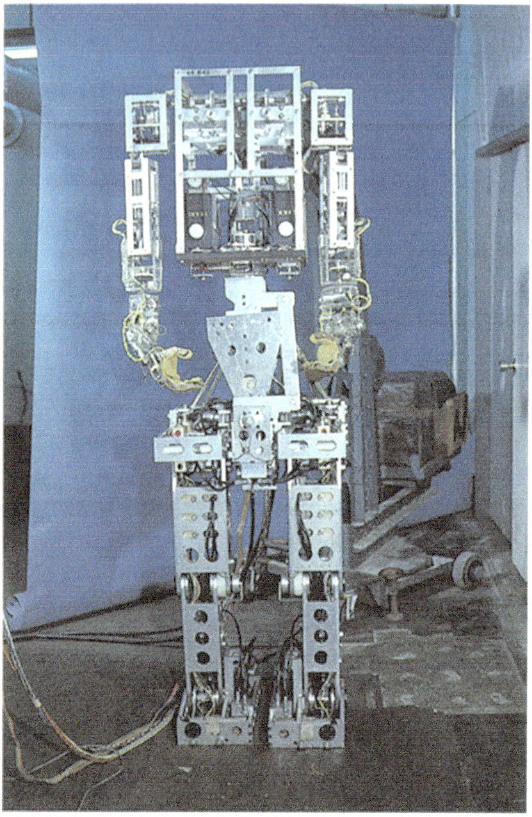

Bibliography

Dirik, M. (2023). The technical ingenuity of Al-Jazari and its relevance to contemporary engineering and design. *Indonesian Journal of Islamic History and Culture*, *4*(1). https://doi.org/10.22373/ijihc.v4i1.2482

Duymaz, Y. K., & Şahin, Ş. (2023). From ancient mythology to modern technology: The historical evolution of artificial intelligence. *European Journal of Therapeutics*, *29*(4), 964–965. https://doi.org/10.58600/eurjther1895

Eaton, M. (2015). *Evolutionary Humanoid Robotics.* Springer.

Fouché, P. J. P. (2025). The legacy of Pierre Jaquet-Droz and his humanoid automata: A Jungian archetypal-interpretivist psychobiography of the writer, draughtsman and musician. *International Review of Psychiatry,* 1–11. https://doi.org/10.1080/09540261.2025.2466504

Frumer, Y. (2020). The short, strange life of the first friendly robot: Japan's Gakutensoku was a giant pneumatic automaton that toured through Asia – until it mysteriously disappeared. *IEEE Spectrum, 57*(6), 42–48. https://doi.org/10.1109/mspec.2020.9099932

Lagrandeur, K. (2012, January 1). *Robots, Moving Statues, and Automata in Ancient Tales and History.* https://www.researchgate.net/publication/350122774_Robots_Moving_Statues_and_Automata_in_Ancient_Tales_and_History

Marsh, A. (2020). The Proto-Roomba - [Past Forward]. *IEEE Spectrum, 57*(3), 50. https://doi.org/10.1109/mspec.2020.9014460

Mayor, A. (2018). *Gods and Robots: Myths, Machines, and Ancient Dreams of Technology.* Princeton University Press.

Meinecke, L., & Voss, L. (2018, July 1). *'I Robot, You Unemployed': Science-Fiction and robotics in the media.* https://www.researchgate.net/publication/324587958_%27I_Robot_You_Unemployed%27_Science-Fiction_and_Robotics_in_the_Media

Palacios-González, C. (2024). Robotic persons and Asimov's three laws of robotics. In *AI Morality* (pp. 185–194). Oxford University Press. https://doi.org/10.1093/oso/9780198876434.003.0018

Stark, L. (2023). Breaking up (with) AI ethics. *American Literature, 95*(2), 365–379. https://doi.org/10.1215/00029831-10575148

Viidalepp, A. (2020). Representations of robots in science fiction film narratives as signifiers of human identity. *Információs Társadalom, 20*(4), 19. https://doi.org/10.22503/inftars.xx.2020.4.2

WABOT-1. (n.d.). Retrieved September 10, 2025, from https://www.humanoid.waseda.ac.jp/booklet/kato_2-j.html

Wiener, N. (2019). *Cybernetics or Control and Communication in the Animal and the Machine.* MIT Press.

Wientzen, T. (2018). Human programming: Brainwashing, automatons, and American unfreedom by Scott Selisker. *Modernism/Modernity, 25*(3), 603–605. https://doi.org/10.1353/mod.2018.0042

CHAPTER

2

From Automation to Intelligence

2.1 The Shift from Basic Automation to Intelligent Systems

Let's take a quick look back. Robots simply repeated before they could walk or speak. The goal of basic automation was to complete a single task flawlessly without ever growing weary or needing a raise. The first real humanoids, such as WABOT-1, which could walk and grasp with all the grace of a broken marionette, were where we left off in the previous chapter. They were puppets, but they were also impressive accomplishments. Every motion was meticulously programmed. Rotate the waist by 30 degrees. Raise your leg 15 centimeters. Put out your arm. Stop.

The era of automation was this one

Imagine a factory floor where a robotic arm welds an automobile door flawlessly, executing a precisely timed ballet of steel and sparks. This is the epitome of automation: exceedingly accurate and relentlessly repetitive. The secret is that the robot is merely obedient; it lacks intelligence. It works in a world that never changes. The illusion breaks the instant a tiny flaw is introduced, like a slight displacement of the car door. The arm doesn't compensate or think. It does what it has been programmed to do, the result being an imperfect weld.

DOI: 10.1201/9788743810704-3

That was acceptable for a while. However, humanoid robots had to leave the factory floor to be anything more than an academic research project. They had to come into our world. Furthermore, our world isn't flawless. It's messy, erratic, and chaotic. Not all of the time is the floor level. A child can leave a toy in the hallway. Someone may abruptly stop in front of you. Not every scenario that could occur in a kitchen can be covered by a single line of code. In a non-ruled environment, a pre-programmed machine is worthless.

Automation changed industries by lowering costs, speeding up production, and decreasing the number of mistakes. But there was a problem: these devices couldn't be changed. You couldn't ask them to do something else or modify; they purely did what you originally requested.

Then things changed

Engineers and researchers began giving machines greater autonomy by equipping them with sensors, vision systems, and, eventually, the capacity to interpret data. As a result, robots were able to perceive their surroundings and respond to changes, such as stopping when a person stepped into their path. This was not intelligence – at least not yet – but it marked a significant improvement over rigid, repetitive behavior. The real turning point, however, came with software, particularly machine learning and AI (Daraghmah & Younis, 2025). These developments transformed robotics by enabling systems to learn from data rather than simply execute predefined instructions. Instead of merely following rules, robots can now anticipate outcomes, identify patterns, and improve their performance over time (Wamba-Taguimdje et al., 2020).

It's like switching from a cassette player to Spotify when you go from automation to intelligence. One plays the song flawlessly every time while adhering to a set track. The other picks up on your actions, comprehends the situation, and modifies its playlist appropriately. Robots must become more like Spotify than like cassette players to function effectively in the real world.

We see this everywhere these days. Robotic assistants in hospitals modify their actions according to the needs of the patients. Intelligent machines maneuver through crowds and changing layouts in warehouses. Vacuum robots in our homes learn from our floor plans and routines. These exist in our everyday lives and are not science fiction. To be clear, though, we're just getting started. Although it is limited, machine intelligence does exist. It is fragile, specific, and restricted. It's unlikely that a robot trained to identify cats would be able to locate a chair. But we're moving forward more quickly than most people think. The interesting part comes when AI is combined with a humanoid form

2.2 Role of Artificial Intelligence and Machine Learning

It's common to conflate AI and machine learning (ML), but it's helpful to see them as distinct yet interconnected concepts. Think of AI as the ultimate goal: making machines intelligent enough to reason, understand, and interact with the world like humans do.

Machine learning is one of the primary methods we use to achieve that goal. Instead of explicitly programming every single rule or answer into a machine, machine learning allows the system to learn from experience, identify patterns, and make predictions or decisions based on data. Let's use a simplistic analogy to illustrate this concept.

Consider teaching a child how to live in the world. AI is like the overarching aim of helping that child navigate and thrive. Machine learning, in this analogy, is akin to teaching the child how to learn – not to memorize facts, but to adapt, understand new situations, and make sense of their experiences as they encounter them. It's about empowering AI to learn from life itself, rather than loading them with every answer. Or we can think of it this way: traditional programming gave a robot a rigid, step-by-step recipe; machine learning gave it the sense of taste.

Instead of providing the machine with exact instructions, we can show it thousands of examples of the desired output and let it figure out how to do it on its own. The process of feeding a system significant amounts of data and allowing it to discover patterns on its own is called machine learning (Mishra, 2024). Everything changed as a result.

For instance, walking may seem easy to us, but it's one of the hardest things for robots to learn. Engineers resorted to machine learning rather than programming every tiny muscle movement. They created virtual worlds in which computer-generated robots could trip, fall, and repeat their mistakes millions of times. Every successful step resulted in a reward, and every failure served as feedback, thanks to a process known as reinforcement learning. The AI gradually gained its sense of balance and stopped simply following a script. It learnt to walk on its own.

Sight underwent the same revolution. You don't write code that defines "handle" or "cylinder" to teach a robot to recognize a "cup." Instead, you feed its AI (a neural network) hundreds of thousands of images labelled "cup." The network analyses this enormous dataset until it discovers the pattern and learns to see a cup in any setting and from any perspective (Srimamilla, 2022).

When teaching a robot to identify something that resembles a cat, the same concept is used. You show it thousands of pictures – large cats, small cats, cats in boxes, cats wearing Halloween costumes – instead of one image and telling it, "This is a cat; memorize it." The neural network develops its internal model of what *makes a cat a cat* after it has seen enough examples to identify the underlying patterns – whiskers, pointy ears, viscous claws and that unmistakable critical gaze (Srimamilla, 2022).

It's math, not magic. However, when it functions properly, it feels magical. The most fascinating part is that AI becomes more intelligent the more data you provide it. Data is the fuel for AI, which has emerged as the new brain. A machine gets smarter the more it can process information from sensors, simulations, or the internet. The race to create an intelligent humanoid has begun.

Robots' capabilities significantly increased once they were able to learn. Moving was no longer the only goal; understanding had become the priority. As the brain finally caught up to the body, significant advances were made in two areas: perception (how robots see the world) and cognition (how they function within it).

The Revolution in Perception: Transitioning from Sensors to Senses

A robot's "vision" before the advent of AI was simply a stream of data from a camera. It didn't know what it was looking at, but it might notice a change in pixels. However, machine learning fundamentally altered this. Robots gained the ability to see thanks to neural networks that were trained on enormous image datasets.

This new ability, known as computer vision, changed everything. Now, a robot can recognize obstacles in addition to avoiding them. It is able to distinguish between a person, a table, and a chair. It is able to read text, identify faces, and decipher gestures. The world transformed from a meaningless jumble of colors and shapes to a well-organized setting full of things and significance. The robot saw a "red ball rolling towards the street" rather than just a red blob (van Eden & Rosman, 2019).

Hearing also improved, going from simple microphones to advanced natural language processing. You can now ask a robot, "Please bring me the book from the table?" and the AI would parse the sentence, understand the intent, identify the key objects ("book" "table"), and create a plan without the need for a specific, pre-programmed command like "EXECUTE WALK SEQUENCE." Perception had evolved into understanding (Younis et al., 2023)

Platforms such as GPT (Generative Pre-trained Transformer) are created in this manner. Large amounts of text, including books, articles, and online discussions, are used to train these sophisticated language models, like GPT-4. By examining this data, the system identifies the rhythms, structures, and patterns of human language, understanding how we ask questions, tell stories, express our opinions, and break down complex concepts. GPT is exceptionally adept at mimicking understanding, predicting the next word based on everything it has previously seen, even though it cannot "understand" in the same way that humans do. That's precisely how it writes essays, conducts conversations, and even contributes to code generation. Imagine then fusing that type of language ability with a humanoid robot that can listen, react, and adjust its facial expressions to fit its tone.

However, seeing the world is only half the battle. The subsequent significant development was cognition, or the capacity for instantaneous thought, planning, and adaptation.

Robots can now think and act in extraordinary ways because they can comprehend their environment. Simultaneous Localization and Mapping (SLAM) is one of the most important innovations. Although it sounds complicated, the concept is straightforward: it is the ability to create a map of an unknown location while also tracking your area on that map. Imagine being dropped into a strange, dark building with only a pencil and paper. You draw the room layout as you go along, and you update a small "You Are Here" dot on the same sketch time after time. This ability, SLAM, allows robots to navigate new and dynamic environments – from a chaotic living room to a disaster zone – on their own without the use of pre-made maps (Garg et al., 2020).

The development of dynamic motion and interaction, enabling robots to move and interact with the physical world, was the next significant step. Robots began to move in surprisingly natural ways due to reinforcement learning, also known as the "school of hard knocks," where behavior is refined through trial and error. They could adjust their steps in real-time while walking on intricate surfaces, such as slippery floors or loose rocks. Instead of falling if pushed, they would stumble and, like humans, attempt to regain their balance. Their hands, which began as simple claws, developed into sensitive tools that could determine the weight and shape of objects, enabling them to pick items up gently without crushing them. They were reacting, adjusting, and responding to the world around them as events transpired, rather than simply adhering to a strict program.

These advances in vision, hearing, movement, and navigation led to the emergence of modern humanoid robots. Robots evolved from simple machines performing predefined actions into complex systems capable of perceiving their environment and adapting to changing conditions. Rather than operating in isolation, contemporary humanoids integrate sensing, control, and learning into a unified whole.

When robots learn on their own, humanity starts to lose control. The trade-off is this. Intelligence makes things both flexible and unpredictable. We'll talk about the risks in more depth later. For now, remember this:

AI and machine learning are what make current humanoid robots possible.

Without them, we would have machines that couldn't move.

As a result, we have machines that can learn, adapt, and make decisions – sometimes in ways that their designers didn't foresee.

2.3 Key Breakthroughs in Humanoid Cognition and Perception

Thus far, we have discussed the evolution of robots from mindless machines to systems with the ability to see, learn, and adapt. Let's get specific now. How does this appear in the actual world? Which humanoid robots truly advanced from simply moving like us to thinking and seeing like us?

Things start to get exciting at this point. Let's look at some of the most important new ideas that changed humanoid robots from stiff machines into stunningly lifelike beings, as well as the robots that showed what was possible.

The theoretical advancements in perception and cognition are abstract concepts until they are actualized in silicon, steel, and wires. Not academic research papers, but tangible machines that move, talk, and interact with our surroundings are what show this shift from automation to intelligence. Each of these trailblazers embodies a unique facet of the quest for AI in a human-like form, transcending mere robotics.

There have been several initiatives that have helped the industry, but four stand out as instances of this progress: Honda's ASIMO, Boston Dynamics' Atlas, Hanson Robotics' Sophia, and Tesla's Optimus. Each machine tells a different story about the problems, aims, and dreams that have shaped the modern humanoid revolution.

ASIMO

ASIMO was there long before films of robots doing backflips became popular. Honda showed off ASIMO (Advanced Step in Innovative Mobility) in 2000 after decades of careful research. For many people, this was the first time they saw a humanoid robot that could be useful in the future. It was a machine made to be graceful, accurate, and able to communicate with people, not rough or strong. ASIMO was a friendly robot diplomat (Hirose & Ogawa, 2006).

ASIMO's most impressive feature was its ability to walk on two legs like a human. ASIMO could walk, run at a reasonable speed (up to 9 km/h in later models), round corners smoothly, and even climb stairs. Earlier robots, on the other hand, shuffled awkwardly. This wasn't caused by a simple pre-programmed routine. It was an early and great use of detecting the surroundings and balancing in real time. Honda's developers came up with "i-WALK" technology, which lets the robot alter its center of gravity in real time, exactly like a person does. If ASIMO started to lean, it would take action to fix the problem. This was the start of the change from rigid automation to flexible adaptation (Shigemi, 2018).

But ASIMO wasn't just a walker. It needed perception to work in places where people lived. It was able to apply computer vision ideas because it had cameras in its head that acted like eyes. It could respond to simple hand movements, follow individuals as they went, and tell who they were by their features. ASIMO would wave back if you waved at it. If you got close enough, it would turn to look at you. This was a very basic type of social awareness that was driven by the AI breakthroughs of the time.

It showed how smart it was by being able to understand regular language. You can offer it simple vocal commands, such "shake my hand" or "walk to this place." It was a big advance, even if it had a small vocabulary and needed to be fed certain words. The exchange was no longer one-sided. The system could understand what people meant and respond in the right way when they said what they wanted. ASIMO also showed that it could do SLAM early on. It could map out its surroundings to get from point A to point B in its well-planned stage shows, avoiding impediments along the way. It could even figure exactly where someone was and how they were standing so that it could bring them a tray without bumping into them.

But ASIMO wasn't smart, it only worked well in places that were organized and easy to predict. The weather had to be just right for its demonstrations. It was more of a well-planned show of what was becoming possible than an entirely separate thing. But its effect can't be overestimated. ASIMO was made for a stage, not for a factory or a war zone. The purpose was to show the world that a humanoid robot might be useful, safe, and friendly. For more than ten years, it was the clear icon of humanoid robots. It was the messenger who brought the cold promise of AI to life and, most crucially, made it easy to understand.

Atlas by Boston Dynamics

Now let's talk about Atlas, Boston Dynamics' humanoid robot. Atlas is about moving quickly, while ASIMO is about strolling carefully.

Atlas knows how to do parkour. Atlas is the experienced Olympian. Atlas, made by Boston Dynamics, a company that has done a lot of research for the military, is always looking for ways to make things stronger and more mobile in the most unpredictable and chaotic situations. Atlas is strong, while ASIMO is calm. It is a real-life demonstration of the most advanced dynamic motion control and reinforcement learning (Kuindersma et al., 2015).

Watching Atlas is a really thrilling experience. Along with walking, it can run, jump over gaps, and throw things with surprising vigor. It does parkour by running up inclined plywood, jumping over obstacles, and ending with backflips and somersaults. This isn't pre-programmed dance moves; it's the result of an AI that has learnt to manipulate its body better than most humans, by doing millions of simulations and real-world tests.

Atlas's intelligence is based on a complex control system that gets data from a variety of sensors, such as cameras, inertial measurement units (IMUs) for balancing, and LiDAR for depth perception. It changes how it moves in real time. At this point, it's very clear that dynamic interaction is happening. If you push Atlas, it will fall over, windmill its arms, and try to get back on its feet. It has a disquieting, animal-like instinct. The AI controller makes the necessary changes in real time, not by using a pre-made animation.

Atlas's brain uses Model Predictive Control to constantly model its next few actions so it can predict the outcome and choose the best course of action. The reinforcement learning algorithms use the important feedback from each failure and misstep to improve their models of motion and physics. Unlike ASIMO, Atlas was not made primarily to interact with people. Its only purpose is to get better at interacting with the outside world. It is a research platform that helps solve the hardest problems in robotics, like moving, manipulating, and lasting in difficult and unpredictable environments. Atlas shows that intelligence is more than simply thinking; it's also about doing.

Sophia

Sophia wanted to learn how to control the face, while Atlas wanted to learn how to control the body. The third and most controversial part of humanoid development is social robotics and emotional AI. Sophia, a product of Hanson Robotics that was turned on in 2016, is an example of this. Sophia was designed to talk, listen, and copy human emotions instead of running or jumping (Parviainen & Coeckelbergh, 2020). The key new thing about it is its face, which is composed of a trademarked material called Frubber (flesh rubber) that looks like skin and is stretched over a complex system of motors to make a wide range of very realistic facial emotions (Faraj et al., n.d.).

The media made Sophia famous. Why? Because she looked like a person. Not only did she look like that, but she also acted like way when she answered, smiled, looked at you, and told jokes, some of which were awkward but nonetheless funny.

Sophia's hybrid intelligence model is made up of a number of different AI technologies. The main reason people like Sophia is because it can talk. Natural language processing software picks up what you say to Sophia and processes it. Like GPT, Sophia was first built on decision trees. Now, it uses powerful cloud-based large language models (LLMs) to make conversations more nuanced, spontaneous, and appropriate to the situation. It can answer questions, make quips, and talk about philosophy.

Its computer vision technology for its eye cameras, which is necessary for creating a sensation of authentic engagement, lets it follow people in a room, recognize faces, and make "eye contact." It pushes the limits of emotional AI by looking at a person's facial expressions and trying to reflect how they feel.

But there is also a lot of debate concerning Sophia. Critics say it's more like an animatronic puppet than a real intelligence since it's a "chatbot with a face" that pretends to be sentient. But the people who made it see it as an artistic character and study platform that looks at how people and robots might interact in the future. Sophia definitely makes us think about the ethics of designing machines that can imitate emotions and what we want from our smart technologies. It makes us establish lines between people and machines, which is what the social frontier is all about.

Tesla Optimus

Now let's talk about Tesla Optimus. When Elon Musk revealed Tesla's humanoid robot in 2021, people weren't sure whether he was serious. But, as is usual for Tesla, things moved quickly. In less than a year, Tesla produced prototypes that could walk, wave, and eventually do simple factory tasks like sorting parts.

But Optimus is special in more ways than just its ability to move and lift things. It's Tesla's brain. The intelligence powering Optimus is a direct extension of the AI that was made for Tesla's self-driving automobiles. This is its main strategic advantage. Millions of Tesla automobiles are collecting lots of information about how people see, understand, and move across complicated human environments. The design of Optimus uses the same exact AI stack. Optimus will be able to move around an office or factory floor using the same neural networks that lets a Tesla car perceive pedestrians, read traffic signs, and predict where other cars will go. It already has a great decision-making and computer vision engine (Shamsuddoha et al., 2025).

Optimus is designed to do much of its thinking "on board," which means that its AI runs locally so it can reply right away without the lag that comes with transferring data to a server far away. This is a strong example of Edge AI. It is required for safe and successful operation around humans. Moving parts, lugging boxes, and doing minor manual labor are some of Optimus's first responsibilities. Its hands are a big part of it since they can hold a lot of different tools and objects. The challenge is in the steady execution of thousands of mundane activities rather than in the attainment of a singular, remarkable achievement. This is a distinct kind of intelligence, one that cares more about generalization (Shamsuddoha et al., 2025).

Figure 2.1 Front view of the humanoid robot Atlas.

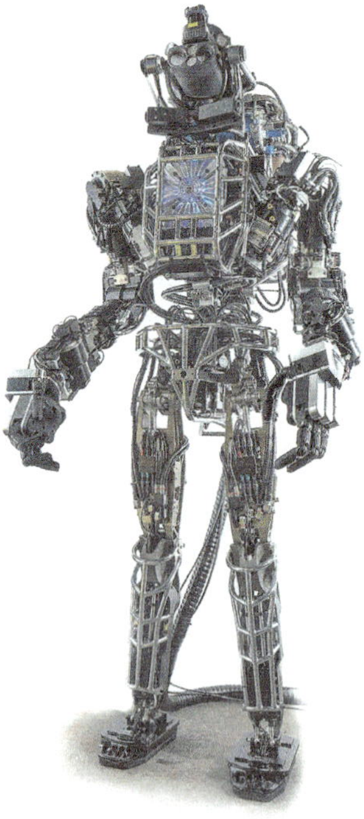

Figure 2.2 ASIMO Humanoid Robot for Rental Business. Source: HONDA, 2001

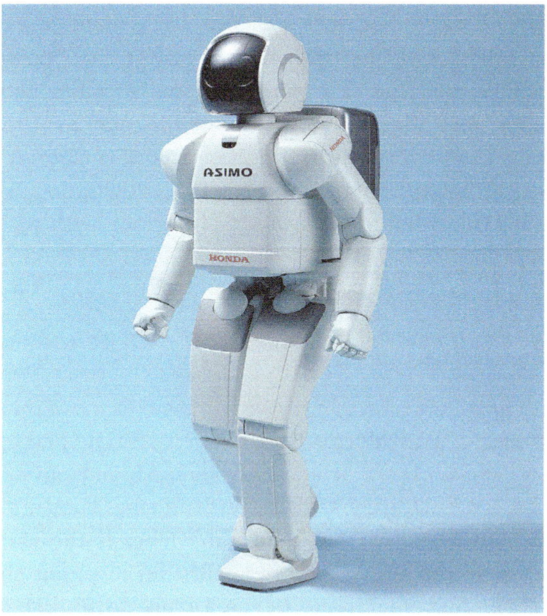

Figure 2.3 Tesla Optimus Gen-2 Humanoid robot

Tesla has a realistic, manufacturing-focused way of doing things. The goal is to build millions of Optimus units while keeping costs low enough that it can compete with human labor. It is a direct descendant of early, single-task automation robots, with one big difference: intelligence. Optimus is meant to be a general-purpose tool that can learn and adapt to new activities through software upgrades, just like a Tesla automobile can get new capabilities overnight. Optimus is still very young, but its purpose is clear: it represents the merging of mass production, robots, and AI. The expectation is that the robot that is the most useful, reliable, and easy to interface with will start the true revolution.

The Trend That is Driving the Changes

So, what do these new ideas have in common? It takes more than simply smart engineering or pretty design. More advanced are real-time perception, adaptive behavior, and the smooth integration of sensors, AI, and physical form. These robots are not just reacting; they are also interpreting. They are made to work, but they can also perceive, learn, and respond. That is the main idea behind how humans perceive and think: to move with purpose, not simply strength; to hear meaning, not just sound; and to see a place to sit, not just a chair. We are not only making powerful machines, but they are also becoming aware of themselves.

A Word of Caution (and Awe)

These new ideas are amazing. But they also bring up difficult questions. If a robot makes a bad choice, who is to blame? What if it learns things you didn't want it to? Once it can read your emotions, does it learn how to control them? In the next chapters, we'll talk about questions like these. But for now, the lesson is that humanoid robots are no longer just prototypes. They are steps forward in progress. They are learning to see, feel, and adapt faster than anyone thought possible. We are already living in a time when machines are smart.

Bibliography

Daraghmah, R. H., & Younis, N. R. (2025). A comprehensive review of machine learning algorithms in autonomous robotics: Challenges and future prospects. *East Journal of Engineering*, *1*(1), 47–61. https://doi.org/10.63496/eje.vol1.iss1.35

Faraj, Z., Morales, C., Hossain, M., Torres, P., & Selamet, M. (n.d.). *Robotic Face Literature Review Group 3 : Green Eggs and Ham Members.* https://api.semanticscholar.org/CorpusID:33581904

Garg, S., Sünderhauf, N., Dayoub, F., Morrison, D., Cosgun, A., Carneiro, G., Wu, Q., Chin, T.-J., Reid, I., Gould, S., Corke, P., & Milford, M. (2020). Semantics for robotic mapping, perception and interaction: A survey. *Foundations and Trends® in Robotics*, *8*(1–2), 1–224. https://doi.org/10.1561/2300000059

Hirose, M., & Ogawa, K. (2006). Honda humanoid robots' development. *Philosophical Transactions of the Royal Society A: Mathematical, Physical and Engineering Sciences*, *365*(1850), 11–19. https://doi.org/10.1098/rsta.2006.1917

Honda (2001, November 12). Honda introduces new ASIMO humanoid robot for rental business. *Honda Motor Co. Ltd.* https://global.honda/en/newsroom/news/2001/c011112-eng.html

Kuindersma, S., Deits, R., Fallon, M., Valenzuela, A., Dai, H., Permenter, F., Koolen, T., Marion, P., & Tedrake, R. (2015). Optimization-based locomotion planning, estimation, and control design for the atlas humanoid robot. *Autonomous Robots*, *40*(3), 429–455. https://doi.org/10.1007/s10514-015-9479-3

Mishra, A. (2024). A comprehensive review of artificial intelligence and machine learning: Concepts, trends, and applications. *International Journal of Scientific Research in Science and Technology*, *11*(5), 126–142. https://doi.org/10.32628/ijsrst2411587

Parviainen, J., & Coeckelbergh, M. (2020). The political choreography of the Sophia robot: Beyond robot rights and citizenship to political performances for the social robotics market. *AI & Society*, *36*(3), 715–724. https://doi.org/10.1007/s00146-020-01104-w

Shamsuddoha, M., Nasir, T., & Fawaaz, M. S. (2025). Humanoid robots like Tesla Optimus and the future of supply chains: Enhancing efficiency, sustainability, and workforce dynamics. *Automation*, *6*(1), 9. https://doi.org/10.3390/automation6010009

Shigemi, S. (2018). ASIMO and humanoid robot research at Honda. In *Humanoid Robotics: A Reference* (pp. 55–90). Springer Netherlands. https://doi.org/10.1007/978-94-007-6046-2_9

Srimamilla, S. (2022). Image classification using convolutional neural networks. *International Journal for Research in Applied Science and Engineering Technology*, *10*(12), 586–591. https://doi.org/10.22214/ijraset.2022.47085

van Eden, B., & Rosman, B. (2019). An overview of robot vision. *2019 Southern African Universities Power Engineering Conference/Robotics and Mechatronics/Pattern Recognition Association of South Africa (SAUPEC/RobMech/PRASA)*, 98–104. https://doi.org/10.1109/robomech.2019.8704781

Wamba-Taguimdje, S.-L., Fosso Wamba, S., Kala Kamdjoug, J. R., & Tchatchouang Wanko, C. E. (2020). Influence of artificial intelligence (AI) on firm performance: The business value of AI-based transformation projects. *Business Process Management Journal*, *26*(7), 1893–1924. https://doi.org/10.1108/bpmj-10-2019-0411

Younis, H. A., Ruhaiyem, N. I. R., Ghaban, W., Gazem, N. A., & Nasser, M. (2023). A systematic literature review on the applications of robots and natural language processing in education. *Electronics*, *12*(13), 2864. https://doi.org/10.3390/electronics12132864

CHAPTER

3

The Titans Behind the Machines

3.1 Strategic Visions: Honda, Boston Dynamics, Hanson Robotics, and Tesla

Chapter 2 talked about the big leaps forward in humanoid abilities, including the discoveries that help machines see, learn, and move better. We examined the technological advancements that enabled the creation of these robots. But the machines we see are more than simply a bunch of clever parts; they are also a reflection of the profound values and long-term ambitions of the people who made them. This chapter changes our focus to the "why" – the main ideas, guiding principles, and unique pathways chosen by the firms that have pushed the limits of humanoid robotics the most. The work of each company conveys a distinct tale about what the ultimate goal of human-like machines in our environment is. Why are Boston Dynamics, Tesla, Honda, and Hanson Robotics racing to build robots that look like people? It's not about showing off. Not really. It's about fixing major problems.

Honda: The Pursuit of Helpful Companionship

Honda isn't trying to build the world's strongest robot. Or the fastest. They're focused on something deceptively simple: everyday life. Their long-running project, ASIMO, wasn't born from military contracts or sci-fi dreams. It came from a question: Can we build a robot that functions in human spaces? (Hirose & Ogawa, 2006). This philosophy shaped every aspect of their research, leading to a robot designed for public service and personal assistance.

DOI: 10.1201/9788743810704-4

Think about your home or office. There are stairs, rugs, uneven floors, and unpredictable people. For a robot, that's a nightmare. It needs to walk, stop, turn, listen, and possibly carry a tray without spilling its contents. Honda believed that if robots were ever going to help people in tangible ways – not just in factories – they had to learn to exist in the same space we do.

Honda's early venture into robotics began in the mid-1980s, driven by a simple yet profound question: Could a machine walk like a person? This curiosity led to the creation of the E-series and P-series prototypes, machines that gradually mastered the complex act of bipedal locomotion. These early robots were clunky and required careful control, but they laid the groundwork for what would become ASIMO, a more advanced robot. The company's thinking was clear: if robots were to operate alongside people in homes and public spaces, they needed to move naturally, without threatening or disrupting human routines. Graceful, fluid movement was not just a technical challenge; it was a fundamental requirement for acceptance.

When ASIMO was finally unveiled in 2000, it embodied this vision. ASIMO was small, friendly-looking, and non-intimidating. Its design was carefully considered to make it approachable, like a character from a children's story. It could walk, run, climb stairs, shake hands, and deliver trays. These abilities were not merely demonstrations of technical prowess; they were expressions of ASIMO's intended role as a social helper. Its real-time balance system, known as i-WALK, permitted it to adjust its posture mid-stride, making its movements remarkably smooth and adaptive compared to its predecessors. This commitment to fluid motion allowed ASIMO to function in human-centered spaces without rigid, pre-set paths.

ASIMO's perception abilities further reinforced Honda's philosophy. It could recognize faces and respond to basic voice commands. These functions allowed for simple, natural interactions, painting a picture of a robot that could serve as a receptionist, a guide in a museum, or an assistant to the elderly. Honda regularly presented ASIMO at public events, schools, and conferences, effectively making it an ambassador for robotics. Its role was to inspire, to show that robots could be safe, friendly, and beneficial members of society. This constant public display was part of Honda's broader communication strategy: to introduce the idea of robots not as threats, but as helpful additions to humanity (Min & Kim, 2024).

Boston Dynamics: Mastering Physical Performance

Boston Dynamics takes a markedly different approach. Their core philosophy is about pushing the very limits of physical ability and control in machines.

For this company, a robot's worth is measured by its capacity to move, balance, and interact with the physical world with unparalleled agility and resilience, especially in complex and unpredictable environments. Their robots are not designed for gentle companionship but for raw athletic performance and the mastery of challenging terrains.

The company's roots in military-funded research, particularly through the Defense Advanced Research Projects Agency (DARPA), heavily influenced this direction. Early projects focused on robots that could navigate rough terrain, carry heavy loads, and operate in dangerous situations where humans could not.

This background instilled a deep commitment to robustness and dynamic stability. Unlike robots designed for clean factory floors or smooth exhibition stages, Boston Dynamics' machines were made for the "school of hard knocks." Their public videos often show engineers pushing, kicking, and challenging their robots, not out of malice, but to gather data on how the machines recover, adapt, and maintain balance under stress. This direct, physical testing is a cornerstone of their development path.

Atlas, their bipedal humanoid, is the clearest expression of this philosophy. Atlas is built to be an athlete. It doesn't purely walk; it runs, jumps over obstacles, performs parkour, and executes complex maneuvers like backflips. These aren't simple pre-programmed actions. They are the result of a control system that utilizes a combination of advanced sensors – including cameras, LiDAR for depth perception, and internal measurement units – to perceive its surroundings and adjust its movements in real-time constantly. The robot's "brain" employs techniques such as Model Predictive Control, where it continually anticipates its next few moves by running simulations to choose the best physical response. This allows Atlas to react immediately to unexpected changes, such as landing on an uneven surface or being shoved. Its ability to regain balance from a severe perturbation, often by flailing its arms and taking corrective steps, is a hallmark of its design (Kuindersma et al., 2015).

Boston Dynamics' path emphasizes learning through repeated trials, including failures. Every stumble provides crucial data for its reinforcement learning algorithms, which refine the robot's understanding of physics and motion. This intense focus on physical mastery means that Atlas, unlike ASIMO, was not initially designed for social interaction. It lacks a friendly demeanor or conversational skills. Its appearance is functional, almost skeletal, underscoring its purpose as a research platform to solve the most difficult problems in physical robotics.

While Atlas remains primarily a research machine, Boston Dynamics has begun to commercialize its technology with robots like Spot (a quadruped) and Stretch (a logistics robot) (Ungureanu et al., 2024). These commercial products demonstrate the practical application of their core philosophy: creating machines that excel at physical tasks in real-world, often messy, settings, from construction sites to warehouses. The company's vision for humanoids is one where machines serve as powerful, agile tools, capable of performing demanding physical labor in environments that are too hazardous or too demanding for people.

Hanson Robotics: The Quest for Social Connection

Hanson Robotics steps into the realm of human-like robots with a distinct focus: building machines that can connect with people through expressive faces and believable personalities. Their philosophy centers on the idea that if robots are to be truly integrated into society, they must be able to understand and display human emotion, engage in natural conversation, and foster a sense of relationship. For David Hanson, the founder, the ultimate goal is to create machines that appear and feel alive, capable of empathy and deep social interaction.

This vision places immense importance on aesthetics and realistic facial expressions. Hanson Robotics developed the patented, skin-like material Frubber, which covers the complex network of motors beneath. This permits their robots, most famously Sophia, to produce a wide array of facial expressions – smiles, frowns, surprise, and even subtle shifts in emotion. The lifelike appearance is central to their strategy, as it's intended to lower barriers to human acceptance and encourage a more natural form of communication.

Sophia, activated in 2016, rapidly became a global phenomenon. She was designed not for physical labor or agility, but for conversation and social engagement. Her "intelligence" is a complex blend of technologies. She uses computer vision to recognize faces, maintain eye contact, and track people's movements. Her auditory sensors capture spoken words, which are then processed by natural language processing software. Critically, Sophia has increasingly integrated powerful cloud-based large LLMs, similar to the technology behind modern chatbots, allowing her to engage in more spontaneous, complex, and contextually relevant discussions. She can answer questions, tell jokes, and participate in philosophical conversations, all while her facial muscles animate expressions that match her words.

Hanson Robotics is at the forefront of what is called "Emotional AI." The aim is to create machines that can not only understand the literal meaning of words but also the emotional cues and subtext. Sophia is a platform for this ongoing exploration. She can analyze a person's facial expression and attempt to mirror their emotional state, creating an illusion of shared feeling. This focus on emotional mirroring and social responsiveness is what distinguishes Hanson Robotics (Faraj et al., 2021).

However, this approach has also generated significant discussion. Critics often point out that Sophia, despite her convincing appearance and conversational abilities, may be more of a sophisticated animatronic puppet than a truly sentient being – chatbot with a face – that creates an illusion of consciousness without proper understanding or feeling. Hanson Robotics acknowledges that Sophia is a research platform and a character, designed to provoke thought and explore the boundaries of human-robot relationships. Her media appearances, including granting Saudi citizenship and addressing the United Nations, were strategic moves to prompt a global conversation about the future of AI and the nature of consciousness. Regardless of whether one sees Sophia as genuinely intelligent or a masterfully crafted illusion, her existence forces us to confront uncomfortable questions about what we want from our intelligent machines: are they tools, or could they become companions? What are the ethical implications of creating machines that can simulate emotions? Hanson Robotics' path is primarily a philosophical and social exploration, pushing us to redefine the boundaries between human and machine.

Tesla: The Drive for Utility and Mass Production

Tesla's entry into the humanoid robotics space with Optimus (also known as the Tesla Bot) represents a pragmatic and utilitarian approach, distinct from that of its competitors. For Elon Musk and Tesla, the purpose of a humanoid robot is clear: to perform practical, repetitive, and often mundane work at a massive scale, addressing labor shortages and boosting productivity. This vision is deeply rooted in Tesla's ambition to create a future of abundance through automation, applying their existing strengths in AI and manufacturing to a new form factor.

The strategic advantage for Optimus lies in its direct connection to Tesla's self-driving car program. The core intelligence powering Optimus is designed to be the same Full Self-Driving (FSD) hardware and neural network that is used in Tesla's autonomous vehicles. This means that Optimus inherits a world-class computer vision and decision-making engine already trained on millions of real-world driving miles. The neural networks that enable a Tesla car to identify pedestrians, navigate complex urban environments, and predict traffic movements are designed to empower Optimus to understand and operate within human environments, such as factories, offices, and, eventually, homes.

Tesla's path for Optimus emphasizes vision-only perception, a controversial but fundamental aspect of their self-driving car strategy. Unlike many other advanced robots that rely on LiDAR or other depth sensors, Optimus is designed to interpret its environment primarily through cameras, much like humans do. This decision streamlines the hardware, potentially reducing cost and complexity for mass production. The robot's intelligence is also designed to operate largely "on-board" (Edge AI), allowing for real-time reactions without constant reliance on cloud connectivity, a crucial aspect for safety and efficiency in practical applications.

The initial tasks envisioned for Optimus are simple and highly practical: carrying boxes, moving parts around a factory, and performing basic manual labor. The focus is not on spectacular acrobatics or deep emotional conversation, but on reliable, repeatable actions. A significant engineering effort is directed towards its hands, designed to be versatile enough to handle a wide variety of tools and objects commonly found in human workspaces. The challenge for Optimus is not to perform a single, impressive feat, but to execute thousands of ordinary tasks consistently and without failure. This points to a different kind of intelligence: one focused on generalization. Can a robot trained to sort items also learn to operate a power tool with minimal retraining?

Tesla's ambition extends beyond just building a functional humanoid; they aim for mass production. Elon Musk has frequently stated the goal of making Optimus affordable, potentially costing less than a car, which would facilitate widespread adoption. This manufacturing-centered approach distinguishes Tesla from the research-focused or niche-market strategies of other players. Optimus represents the idea that the true revolution in robotics will come when humanoids become cost-effective, general-purpose tools, capable of learning and adapting to new tasks through software updates, much like Tesla cars receive new features wirelessly (Husain, 2024). While still in its early stages, Optimus's vision is clear: to bring adaptive, intelligent machines to the workforce at a scale that could reshape global industries. It is a pragmatic gamble that the ultimate purpose of humanoids lies in their utility and widespread availability.

3.2 Comparing Design Philosophies and R&D Approaches

Hard vs. Soft Robotics

One of the most fundamental distinctions in the creation of humanoid robots lies in the choice of materials and construction methods. This divides the field broadly into "hard" robotics and "soft" robotics, each bringing its strengths and limitations to the fore.

The "hard" approach to robotics prioritizes strength, precise movement, and the ability to handle significant physical force. Robots built with this philosophy typically employ rigid components, often made of aluminum or titanium, and hard plastics. Their joints are powered by strong electric motors or hydraulic systems, allowing for controlled, powerful, and often rapid movements. The core idea is to create machines that can exert considerable force, move with great accuracy, and withstand harsh conditions (Alici, 2018).

Boston Dynamics' Atlas serves as a prime example of complex robotics. Its skeletal frame, visible joints, and reliance on hydraulics (in earlier versions, now electric motors) make its dedication to physical capability clear. The company's work with Atlas is about pushing physical boundaries. Their research approach involves extensive computer simulations to predict how the robot will react to different forces and situations. They then test these predictions in the real world through intense physical challenges, observing how Atlas moves, balances, and recovers (Guizzo, 2019). The data collected from these trials, even from falls or stumbles, is crucial for refining the robot's control systems. The goal is to build machines that can run, jump, lift, and navigate challenging places with unmatched athleticism. This approach results in robots that are highly capable in physical tasks but, by their nature, can be less forgiving in direct contact with humans due to their power and rigid structures. Their sound, often a whirring of motors or the hiss of hydraulics, also marks their presence.

The benefit of complex robotics is the sheer performance it permits. Such robots can carry heavy loads, move at impressive speeds, and maintain balance in dynamic situations. They are designed to operate in factory settings, construction sites, or disaster zones, where strength and endurance are paramount. The downside, however, can be their lack of inherent safety when operating closely with people. Their precise, powerful movements, while efficient for tasks, may not be as naturally compliant or gentle for close human interaction. Any accidental contact can result in significant force (Kuindersma et al., 2015).

In contrast, soft robotics aims for qualities such as compliance, safety in human contact, and fluid, lifelike movements. This approach draws inspiration from biological systems, which often rely on flexible materials rather than rigid bones and joints. Soft robots are constructed using pliable materials like silicone, rubber, or flexible plastics, sometimes combined with fluid-driven actuators (like pneumatic or hydraulic systems using low pressure) or innovative materials that change shape when stimulated (Alici, 2018).

Hanson Robotics' Sophia embodies many principles of soft robotics, particularly in its outer layers. While Sophia has rigid internal structures for movement and support, its patented Frubber skin and underlying network of small motors create a remarkably expressive and human-like face. This skin is soft and pliable, allowing for subtle changes in expression that convey a wide range of emotions. The emphasis here is not on brute strength or rapid locomotion, but on creating a visually and tactilely appealing appearance that encourages human connection. The soft materials contribute to a perception of approachability and reduce the feeling of threat (Karmakar & Sarkar, 2019).

The research approach in soft robotics often involves deep exploration into materials science and bio-inspired design. Engineers study how biological bodies move and interact with their surroundings, seeking to replicate those qualities in artificial systems. For soft robots designed to interact with humans, safety is a primary concern, leading to designs that can absorb impact or deform without causing injury. These robots tend to move more slowly and gently, with a focus on natural, flowing gestures rather than swift, forceful actions.

The primary benefit of soft robotics is the potential for safer and more natural interactions with people. Their compliant nature allows them to deform upon contact, thereby reducing the risk of harm. They can also appear more "friendly" and less machine-like. However, soft robots typically have limitations in terms of load-bearing capacity, speed, and precision for heavy industrial tasks. They are better suited for tasks that require light manipulation, social interaction, or applications where flexibility and adaptability to irregular shapes are more important than brute force (Polygerinos et al., 2017).

It is important to note that the division between hard and soft robotics is not always absolute. Some modern robots blend elements of both, using rigid structures for power where needed and soft materials for surfaces that interact directly with people. However, the underlying design philosophy – whether to prioritize strength and precision or compliance and gentle interaction – remains a core distinction that guides their creation.

R&D Approaches

The design philosophies we have discussed are deeply intertwined with the research and development (R&D) approaches companies choose to pursue. These approaches can generally be grouped into categories ranging from pure scientific inquiry to product creation on a large scale.

Some organizations, particularly academic institutions and the early phases of companies like Honda and Boston Dynamics, operate with a purely research-driven R&D approach. The primary goal here is to push the fundamental limits of what is possible in robotics. This means exploring new control algorithms, novel materials, advanced sensor technologies, or entirely new ways for robots to learn and perceive.

The output of such research is often scientific papers, laboratory prototypes, and conceptual demonstrations. There is less immediate concern for commercial viability, cost-effectiveness, or immediate application. Funding typically comes from grants (like those from DARPA for Boston Dynamics) or internal corporate R&D budgets dedicated to long-term discovery. The aim is to solve complex scientific puzzles, such as stable bipedal walking over uneven ground (as Honda pioneered) or dynamic balancing during extreme maneuvers (as Boston Dynamics showcases with Atlas). This approach builds the foundational knowledge that later commercial products can stand upon.

Once basic scientific problems are partially solved, many companies shift to a problem-solving or application-driven R&D approach. Here, the focus moves from "what's possible?" to "what problem can this robot solve?" The goal is to design and refine robots for specific real-world needs.

This includes creating specialized robots for specified industries, such as logistics (Boston Dynamics' Stretch, designed for moving boxes in warehouses), healthcare (surgical robots or assistive devices), or disaster response. The R&D process becomes more focused on user requirements, environmental adaptability for a specific task, and making the robot sufficiently robust for its intended work. Commercial contracts, venture capital, and strategic partnerships often fund this stage. The output is a functional robot that can perform a defined set of tasks reliably in a specific setting. While not necessarily mass-produced, these robots are built with practical deployment in mind.

Mass Production and Scalability-Driven R&D

The most ambitious R&D approach, exemplified by Tesla with Optimus, is driven by the desire for mass production and broad market availability. Here, the robot is conceived as a product from its inception, with manufacturability, cost reduction, and widespread adoption as central tenets.

This approach requires a different mindset. Design choices are heavily influenced by the need for simplified components, automated assembly processes, and a streamlined supply chain. Performance must be balanced with affordability. Tesla's decision to base Optimus's intelligence on its self-driving car AI is a prime example of this: it permits them to reuse a well-tested, camera-centered system and potentially reduce the need for expensive, specialized sensors, making the robot more suitable for large-scale production. The R&D process involves rapid iteration, designing for manufacturing ease, and extensive testing for long-term reliability in diverse conditions, not just laboratory settings. Funding often comes from public markets or anticipated sales revenue. The goal is not just to build one robot, but to build millions, making humanoid robotics accessible and potentially omnipresent in daily life.

3.3 Ethics and Public Perception

Let's face it: making machines that look and act like us is more than just a puzzle for engineers. It opens up a whole world of moral questions, human responses, and outcomes we cannot always foresee. When a machine smiles at you, speaks your language, and reacts to your touch, it stops feeling like a simple device. It starts to feel like something more, and that feeling changes everything about how we think about it.

Humanoid robots have stepped out of the controlled laboratory environment and into our daily lives. This means people are watching them closely. They are forming opinions. They are reacting in various ways. They are asking questions. Some of these questions are straightforward. Is this robot safe to be near? Can it do its work well? Other questions are much more complex to answer. Is this machine pretending to care about me? Should I treat it as if it were a person?

Robots are no longer just tools. They interact with children, assist nurses, greet visitors in hotels, and appear on television programs. They appear in popular online videos, news reports, and classrooms in schools. Each time someone sees one, they bring their own beliefs, their hopes, and their worries along with them.

The Unsettling Effect

Let's discuss a typical reaction: humanoid robots can sometimes evoke a sense of unease. This phenomenon is often referred to as the "uncanny valley." It describes what happens when something resembles a human, but not quite. Think of a figure made of wax that moves slightly. Or a face that blinks just a bit too slowly. Instead of feeling comfortable, we become uneasy. Our minds struggle to decide whether the thing we are observing is truly alive or not, and the result is a sense of discomfort.

Some robots fall deeply into this unsettling zone. Sophia, for example, has sparked both great interest and a degree of unease due to the human-like nature of her facial expressions. People respond strongly to her. Some are amazed by her abilities. Others experience something closer to a feeling of dread. This is not a fault in the robot itself; it is a clear sign of how complex our human responses can be.

And it's not just about how a robot appears. It also has to do with how it moves, how it speaks, and even how it stands still. A robot that pauses for too long before answering, or stares for just a second too long, can create discomfort. We are accustomed to people behaving in particular ways, and when a robot imitates that behavior closely but without fully matching it, we are thrown off balance.

Designers now spend many hours trying to strike the right balance between familiarity and unease. Some choose a more stylized appearance – not entirely human, but close enough to seem friendly. Others lean into a more mechanical look, making it clear that the machine is not human. The aim is to avoid the unsettling effect. The goal is to create something that feels secure, easy to relate to, and artificial.

Trust and Expectation

There is an interesting aspect to humanoids: we often expect more from them than is reasonable. When something has a face, a voice, and a body, we tend to assume it understands what it is doing. We give it credit for being intelligent. We place our trust in it. This can be risky.

In some studies, people place more trust in robots with human-like features than they do in mechanical ones, even when the mechanical robots perform more reliably. This presents a serious concern, particularly in settings such as medical centers, travel hubs, or residential care homes. A machine that looks friendly might be given duties it is not ready for, simply because people believe it can handle them (van Pinxteren et al., 2019).

The issue runs deeper. Some people form emotional connections with these machines. They talk to them. They share personal thoughts with them. In some settings for older people, residents have been observed embracing and thanking the robots assigned to help them. This might seem heartwarming, but it raises a challenging point. What happens when people begin treating machines as if they were friends? (Finkel & Krämer, 2022).

Is this fair to the human? What if they genuinely believe the robot cares about them? What if they share private thoughts, hoping for understanding in return? Is it right to permit such an illusion?

There is also the problem of placing too much faith in it. A robot might excel in one task and then be assumed to be capable in other areas. A robot that can lift a box might be tasked with caring for children. A robot that can hold a conversation might be expected to keep secrets or offer advice. We blur the lines between machine and companion more quickly than we realize.

Identity, Emotion, and Consent

Let us consider another point. Imagine a robot that looks like someone you know, a well-known public figure, or someone who never gave their consent for their likeness to be used. This is not just something from stories. It is already happening.

In some places, humanoid robots are being designed to resemble real people. Some companies have attempted to replicate the faces and voices of famous individuals, as well as family members who have passed away, for personal robots. The technology exists to create a version of someone who is no longer present – and have it smile at you, talk to you, and look you in the eye (Jee, 2022).

This raises a complex area of moral questions. Who has right to a face? Who has right to a voice? If you pass away, can your family allow a robotics company to use your likeness? Can a corporation do so? What happens when someone sees a robot version of their friend walking in public?

Then there are feelings. Many modern humanoids are being taught to recognize human emotions. They can detect sadness in your voice or notice when your eyes are watering. They respond to emotional signals. Some even try to offer comfort. But the robot itself does not feel anything. It is imitating empathy. Is that comforting, or is it a form of manipulation?

Critics argue that an AI that mimics emotion creates a false sense of connection. You might feel cared for, but the robot has no true feelings of its own. Others argue that as long as the outcome is beneficial – such as calming a child or reducing loneliness in an older person – then the imitation is acceptable.

But that boundary is hard to define. Especially when we begin using these machines in sensitive environments: medical centers, schools, and homes. Relationships formed with machines can become complex very quickly.

Then there is the matter of permission. If a robot records your face, your voice, and your actions, what happens to that information? Are you told it is being collected? Can you say no to it? If it is gathering information while you talk to it and using that information to improve its software, are you part of a test you never agreed to?

These are not just technical details. They are questions about human dignity, private life, and trust.

Cultural Bias and Representation

Humanoid robots do not exist in isolation. They show the influence of the cultures that built them.

Many of the most well-known humanoids have been designed in specific parts of the world – Japan, South Korea, and the United States. As a result, their appearances, body language, and voices often mirror those cultural standards. The way a robot bows in Japan might be seen as polite there but is unusual elsewhere. The way a service robot speaks in Korea might come across as impolite in a different culture.

This can cause robots to carry cultural biases unintentionally. For example, if a robot's standard appearance is light-skinned and Western, what message does that send to a global audience? If its "normal" behavior is based on one culture's idea of politeness, how does it act in another?

Even more concerning is the information used to train these machines. Language models, facial recognition systems, and emotional detection tools are often built on datasets that do not capture the full range of human experiences. This means a robot might be good at understanding the expressions of one group but fail entirely with another. It might respond correctly to one accent but struggle with others.

The outcome? Could some people receive better treatment from robots than others? Not because of bad intentions, but because of their design. This is a significant moral gap that engineers and companies must address if humanoid robots are to be deployed worldwide.

People are often afraid of what they do not understand. And when it comes to humanoid robots, many people still do not fully grasp them. Popular stories have not helped. From machines that kill in movies to robots replacing human workers, popular culture has instilled strong emotional ideas in people's minds. These stories matter. They shape how people react to real machines.

When a humanoid robot appears in public, people bring all their worries and their fantasies with them. Some are excited. Others are anxious. Some groups are hostile. There have been documented instances of people damaging or attacking robots in public places, not because the machines did anything wrong, but because their presence made people uncomfortable.

At the same time, there is a risk of overstating what these machines can actually do. Companies sometimes exaggerate the intelligence or independence of their humanoids, causing people to believe they are far more advanced than they truly are. This creates false hopes and, eventually, disappointment.

It also creates chances for people with bad intentions. Technology that creates realistic fake images or sounds, combined with humanoid robots, could one day be used to simulate the appearance and behavior of real people. Campaigns spreading false information could utilize robotic figures to convey misleading messages. The more lifelike these machines become, the simpler it is to blur the line between what is real and what is not.

Regulation and Responsibility

Currently, there are very few regulations governing what humanoid robots can and cannot do. Governments are just beginning to legislate. Some are suggesting laws regarding the collection of information and AI that mimics emotions. Others are focusing on product safety and who is responsible if something goes wrong. However, most countries do not have specific rules regarding robots that resemble humans. Who is responsible when a humanoid robot causes harm? Is it the maker? The person who wrote the computer code? The person who turned it on? If a robot gives bad advice or records something it should not, who is to blame? If it stops working correctly and someone gets hurt, what happens next?

These questions are no longer just ideas; they are now realities. As humanoids move into workplaces, public areas, and private homes, the need for clear accountability is pressing. Some experts have suggested a kind of "robot charter" to define what machines can and cannot be designed to do. Others argue that the focus should be on human responsibility, not on machine rights. Either way, something must change. The world needs legal and ethical guidelines to prevent technology from moving too far ahead of public safety.

What People Desire

Despite all the difficulties, one thing is clear: people care about this. They might be doubtful, curious, afraid, or fascinated, but they are paying attention. This means public perception is not just a minor point. It is central to how humanoids will be accepted, used, and trusted. People want robots to help, not to mislead. They want honesty about what the machines can do. They want control over their personal information. They want machines that are secure, open about their workings, and show respect for human worth.

Most of all, they want to feel that they matter more than the machine. Humanoid robots are not just reflections of our technology. They are reflections of ourselves. How we design them, how we treat them, and how we react to them shows what we value – and what we fear.

In the following chapters, we will look at how these matters play out in everyday places: medical centers, schools, offices, and even our own homes. But as we move forward, we cannot ignore what we have already seen. Creating robots that resemble us means confronting everything that makes us human – the good, the challenging, and the complex.

Bibliography

Alici, G. (2018). Softer is harder: What differentiates soft robotics from hard robotics? *MRS Advances*, *3*(28), 1557–1568. https://doi.org/10.1557/adv.2018.159

Faraj, Z., Selamet, M., Morales, C., Torres, P., Hossain, M., Chen, B., & Lipson, H. (2021). Facially expressive humanoid robotic face. *HardwareX*, *9*, e00117. https://doi.org/10.1016/j.ohx.2020.e00117

Finkel, M., & Krämer, N. C. (2022). Humanoid robots – artificial. Human-like. Credible? Empirical comparisons of source credibility attributions between humans, humanoid robots, and non-human-like devices. *International Journal of Social Robotics*, *14*(6), 1397–1411. https://doi.org/10.1007/s12369-022-00879-w

Guizzo, E. (2019). By leaps and bounds: An exclusive look at how Boston dynamics is redefining robot agility. *IEEE Spectrum*, *56*(12), 34–39. https://doi.org/10.1109/mspec.2019.8913831

Hirose, M., & Ogawa, K. (2006). Honda humanoid robots' development. *Philosophical Transactions of the Royal Society A: Mathematical, Physical and Engineering Sciences*, *365*(1850), 11–19. https://doi.org/10.1098/rsta.2006.1917

Husain, A. (2024, November 4). Humanoid robotics will transform manufacturing and the global economy. *Forbes*. https://www.forbes.com/sites/amirhusain/2024/11/04/humanoid-robotics-will-transform-manufacturing-and-the-global-economy/

Jee, C. (2022, October 18). Technology that lets us "speak" to our dead relatives has arrived. Are we ready? *MIT Technology Review*. https://www.technologyreview.com/2022/10/18/1061320/digital-clones-of-dead-people/

Karmakar, S., & Sarkar, A. (2019). Design and implementation of bio-inspired soft robotic grippers. *Proceedings of the Advances in Robotics 2019*, 1–6. https://doi.org/10.1145/3352593.3352618

Kuindersma, S., Deits, R., Fallon, M., Valenzuela, A., Dai, H., Permenter, F., Koolen, T., Marion, P., & Tedrake, R. (2015). Optimization-based locomotion planning, estimation, and control design for the atlas humanoid robot. *Autonomous Robots*, *40*(3), 429–455. https://doi.org/10.1007/s10514-015-9479-3

Min, C.-H., & Kim, Y. M. (2024). ASIMO: Agent-centric scene representation in multi-object manipulation. *The International Journal of Robotics Research*, *44*(1), 22–64. https://doi.org/10.1177/02783649241257537

Polygerinos, P., Correll, N., Morin, S. A., Mosadegh, B., Onal, C. D., Petersen, K., Cianchetti, M., Tolley, M. T., & Shepherd, R. F. (2017). Soft robotics: Review of fluid-driven intrinsically soft devices; manufacturing, sensing, control, and applications in human-robot interaction. *Advanced Engineering Materials*, *19*(12). https://doi.org/10.1002/adem.201700016

Ungureanu, L. M., Munteanu, I. S., & Vulturescu, V. (2024). Recent advances and industrial applications of the synergy between robots and artificial intelligence globally in the era of industry 4.0. *MDPI AG*. https://doi.org/10.20944/preprints202412.2535.v1

van Pinxteren, M. M. E., Wetzels, R. W. H., Rüger, J., Pluymaekers, M., & Wetzels, M. (2019). Trust in humanoid robots: Implications for services marketing. *Journal of Services Marketing*, *33*(4), 507–518. https://doi.org/10.1108/jsm-01-2018-0045

Part II

Automation and Human Interaction

CHAPTER

4

Humanoids in the Workforce

As humanoid robots become more capable, moving beyond laboratory demonstrations and into practical settings, a significant area of discussion centers on their entry into the workforce. For centuries, humans have sought ways to automate labor, first with simple machines, then with dedicated industrial robots. Now, with machines that walk, grasp, and communicate in ways that are familiar to us, the scope of automated tasks is expanding rapidly. This chapter begins our exploration of where these human-like machines are finding their place, focusing on their potential functions across various important sectors: manufacturing, logistics, healthcare, and various service industries. The core idea here is not just replacing human effort, but bringing adaptable, sometimes versatile, forms of machine assistance to places designed for human activity.

4.1 Roles in Manufacturing, Logistics, Healthcare, and Service Industries

Automation has been a part of manufacturing for a long time. For decades, huge, fixed robotic arms have been used to weld, paint, and transport heavy objects in special areas, usually behind safety gates. These antiquated machines are great at doing the same exact thing over and over again. They are strong and never get tired, but they can't change. These robots generally need a lot of retooling when the product changes or the work environment changes, which is expensive and takes a lot of time.

Humanoid robots are a different kind of help for the industrial floor. Their main benefit is that they look like people. They can utilize tools and live in

DOI: 10.1201/9788743810704-6

places intended for people since their body layout is similar to ours: they have two arms, two legs, and hands that can grab. This means that a new machine doesn't need to be created from scratch in a factory. A humanoid can stroll down existing aisles, stand at workstation made for a person, then pick up a wrench or screwdriver without needing to change the shape of the tool. They can do a lot more than industrial arms since they can adapt to the environment.

Think about the assembly line. Many processes need precise manual operations that are hard for old-fashioned machines to do, or they need decisions that are easier for a person to make. As humanoids get better at using their hands, they can start to do these things. They can pick up small electronic bits and carefully put them on a circuit board. They can use their vision sensors to check final products for faults, join fasteners, or attach wire harnesses. They can do the kinds of things that people frequently find tedious or tiring since they are repetitive, precise, and sometimes boring.

Picture a humanoid, maybe Tesla's Optimus, working in a factory that makes cars. This machine could walk to a workstation where a person used to stand instead of needing a whole new robotic cell designed just for it. It can pick up a door panel, use some regular tools to bolt it down, and then move on to the next station. It can work with machines that already have buttons, levers, or screens that are designed for people to use. One of the best reasons to use humanoids is that they fit into existing workspaces without needing major adjustments to the layout. These kinds of machines can help load and unload materials onto production lines, move finished goods, or even do quality control by comparing products to digital models. The idea is to get people out of the most physically demanding or boring jobs so they can focus on overseeing, solving problems, or more complicated duties that still need a human touch.

Logistics and Warehousing

There are always a lot of things moving in the area of logistics, which includes warehouses, distribution centers, and shipping hubs. Automated Guided Vehicles (AGVs) are wheeled robots that move along fixed tracks, while Automated Storage and Retrieval Systems (AS/RS) are complicated systems that store and retrieve vast amounts of the same type of commodities (Patricio & Mendes, 2020). These systems work very well for transferring large amounts of stuff or keeping track of flows that are easy to forecast. But they have trouble with things that aren't regular, situations that aren't structured, and the wide range of goods that are available in today's online stores.

Humanoid robots add a whole new level of capabilities to this bustling field. One big benefit is that they can move around in environments that aren't structured. There are typically obstructions, limited lanes, and floors that are different in warehouses. A humanoid can walk around a pallet that is in the wrong position, step over a cable, or get through busy spaces in ways that wheeled robots can't. This mobility lets them traverse more ground and get to things that are harder to get to.

Picking and packing could be one of the most useful things humanoids could do in logistics. Modern warehouses handle millions of items, and each one is different in shape, size, and weight. At the moment, robotic arms that are used for picking can only pick up certain types of items. Humanoids can pick up a number of things, from soft clothes to hard electronics, thanks to their multi-jointed arms and flexible hands. They can reach high shelves, bend down to pick things up off the floor, and gently place them in boxes. Because they can touch different things and find their way around shelves, they are perfect for order fulfilment, which involves gathering a wide range of products for each customer's purchase (Garabini et al., 2021).

Humanoids can do more than pick and pack. They can help organize shipments, load and unload vehicles or containers, and keep track of stock. They may transfer things from the ports where they were received to storage facilities, or from packing stations to shipping zones. They could use their vision sensors to monitor inventories by counting goods or checking product codes. Picture a humanoid moving quickly but carefully around a huge e-commerce fulfilment center. It doesn't have to stay on a track; it can vary its path as needed. It picks out specific items for customer orders and places them in boxes. This kind of work is quite hard on the human body and requires a lot of people, thus it's a great place for humanoids to help.

Healthcare and Support Services in Assisting Caregivers and Patients

There are a lot of obstacles and a lot of potential for humanoid robots in the healthcare field, where they need to be attentive, sanitary, and aid patients. Some specialized robots already assist with surgery or deliver medical supplies, but humanoids promise to play a bigger role by assisting medical professionals and directly helping patients (Mukherjee et al., 2022; Rasmussen et al., 2024).

One of the best things about humanoids in healthcare is that they can help with patient care in ways that are physically difficult for human professionals to do. This includes things like helping patients move from a bed to a chair, aiding them with walking rehabilitation, or holding someone up as they stand. Because humanoids look like people, they can use specialized lifting equipment and patient aids that are made for people. Nurses and other carers often have to lift heavy things (Ozturkcan & Merdin-Uygur, 2021). Humanoids could help nurses and other carers avoid injury by taking up the physical work of moving patients.

Humanoids could aid with more than just physical tasks; they could also help with monitoring and support. With sensors, they may check basic vital signs, keep a record of how patients move, or remember to give drugs. Their presence could also be company for elders or people in long-term care facilities. Humanoids with social skills (like those in Hanson Robotics' designs) could have simple discussions with patients, read them stories, or remind them of appointments. They couldn't feel emotions, but they would be a reassuring presence and help patients feel less alone.

Also, humanoids could do many cleaning and maintenance jobs in hospitals and clinics. This includes cleaning patient rooms, sterilizing equipment (if the hospital has the right tools), and moving linens and supplies between different parts of the hospital. Humanoids might handle delicate samples, count pills, or make batches of medicine in a pharmacy or lab. This would make sure that everything was done accurately and let technicians do more important tasks (Tanioka et al., 2019). Think about a humanoid carefully walking through the hospital halls, bringing a patient their meal or helping a nurse by getting a specific supply item from a storage cupboard. Humanoids could take care of these repetitious, logistical, or exhausting chores, which would free up more of the medical staff's time and energy to deal with patients directly and make complex medical choices where their unique human skills are irreplaceable.

Service Industries in Enhancing Public Experiences and Daily Tasks

There are many different kinds of businesses in the service sector, such as retail, hospitality, education, and public safety. In these industries, it's normal to connect with people directly, frequently in less formal settings. Automated technologies that are now in use include self-service kiosks, rudimentary delivery robots on wheels, and automated check-in systems. Humanoid robots can walk around freely and engage in more natural ways, which opens up new opportunities for more personalized and flexible help.

Humanoids might greet customers in stores, show them where to find requested items, or give them basic information about stock. They could help with stacking shelves, arranging merchandise displays, or getting things from stockrooms because they have nimble hands. Think of a humanoid in a department store who is helping a customer find something or carrying a basket of items to the register. This could allow salespeople to give a more personalized service or deal with more complicated consumer questions (Barakeh et al., 2019).

Humanoids could work as concierges in hotels, giving guests information about local attractions and hotel services. They could deliver room service, carry luggage, or clean public spaces like hallways and lobbies. A humanoid might check people in or show them to their rooms in a hotel (Zhong & Verma, 2022). These interactions feel more natural than dealing with a screen or a boxy machine since the shape of the robot makes them feel more like a person. This could make the guest experience more welcoming.

Humanoids could help teachers in a number of ways in the classroom. They could lead simple exercises, give one-on-one instruction in basic areas like math or reading, or get kids involved in interactive learning activities. A humanoid may tell stories or lead simple group exercises for younger learners. This would make the learning process more fun and interesting. They wouldn't take the role of human teachers, but they may be helpful tools that provide students repetitive lessons or help with problems that let teachers focus on each student's needs and more difficult subjects (Brown & Howard, 2013).

Lastly, humanoids could keep surveillance on public spaces and enforce safety by responding to basic warnings, keeping an eye on sports fields, or helping out in crowded situations. A humanoid with good navigation skills may patrol a park or stadium, looking for problems or helping people find their way to safety exits. Their presence could also stop people from doing things they shouldn't (Younis et al., 2021).

The fact that humanoid robots can work in manufacturing, shipping, healthcare, and service industries shows one of their main benefits: they can bring adaptive automation to places that are made for people. Humanoids are more flexible than specialized robots because they can use existing equipment, move across complicated locations, and do a wide range of tasks that need a human-like shape. Even while many of these applications are still in their early phases, the research and development behind them are pointing to a future where humanoids will be able to help us with a wide range of tasks that need mobility, dexterity, and interaction with our surroundings. The next sections will go into more detail on the pros and cons of these devices moving from theory to reality.

Figure 4.1 A humanoid robot performing tasks in a warehouse and a hospital, demonstrating industrial and medical applications.

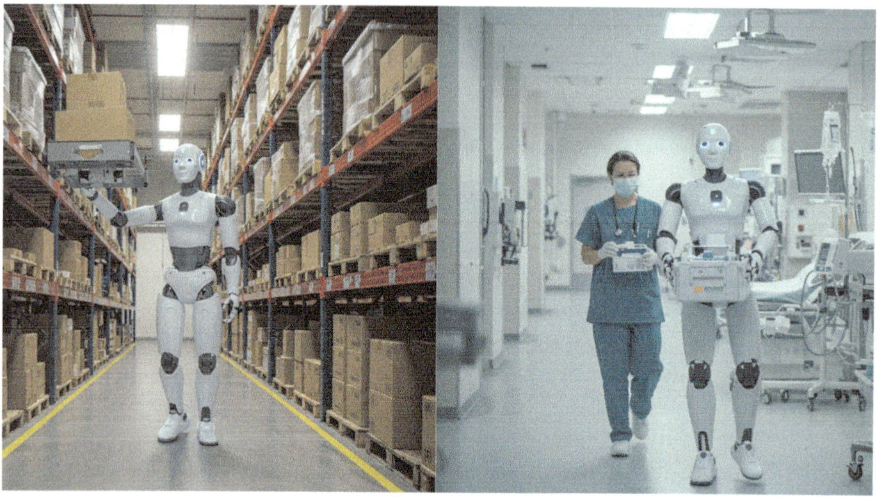

4.2 Benefits and Limitations

In Section 4.1, we explored the various ways in which humanoid robots are beginning to find their place in our work environments, moving into factories, warehouses, healthcare settings, and service roles. We discussed how their human-like shape enables them to fit into workspaces and use tools designed for people. Now, we will examine the practical advantages these machines offer when they join a work team, as well as the significant hurdles and drawbacks that accompany their presence. Understanding both the positive aspects and the challenging aspects is essential for a comprehensive picture of their likely effects.

The Good Points: Why Humanoids are Joining the Work Team

The drive to bring humanoid robots into work settings is impelled by several clear advantages they offer compared to human workers or even traditional machines that are less adaptable. These benefits often focus on working more effectively, making places safer, and handling tasks that are difficult or not enjoyed by employees.

One of the most beneficial aspects of humanoids, like any automated system, is their ability to maintain constant activity. Unlike people, robots do not need breaks, sleep, or compassionate leave. They can operate around the clock, every day of the week, every week of the year, if required, significantly increasing the total amount of work done in a factory, storage facility, or service provider's location. This steady effort can lead to substantially higher production rates and faster completion of tasks. Beyond working long hours, humanoids can perform tasks with great steadiness. Once set for a particular action, they can perform it precisely thousands or millions of times without getting tired or losing focus. This leads to a more uniform product or service quality, and less mistakes and wasted materials that can occur when humans are tired or not paying attention. If making things, every part is put together in the same way. In moving goods, every package is handled with the same care. This dependable, unchanging performance is a significant reason why businesses want to utilize robots to enhance their overall product and make it more consistent.

The most widely accepted reason for using robots is their ability to perform tasks that are dull, dirty, or dangerous. These are often referred to as "3D" jobs (Owen-Hill, 2022). Many industrial settings pose dangers from heavy equipment, hazardous materials, extreme temperatures, or unstable structures. Humanoid robots can be sent into these places without putting human lives at risk. For example, when disaster strikes, a humanoid could enter a building that has collapsed to search for people, remove debris, or assess the extent of the damage, duties too hazardous for human rescue teams. In factories, workers could perform welding in hot, confined spaces or work with dangerous substances. Other jobs involve being exposed to dirt, dust, or unclean places that are unpleasant or unhealthy for human workers. Humanoids can operate in these surroundings without complaint, and their surfaces can be easily cleaned or made germ-free. This is especially important in areas that require extremely high levels of cleanliness, such as medical facilities or food processing buildings, where a human presence can introduce unwanted contaminants. Lastly, many duties are repetitive and unexciting (Royakkers & van Est, 2015). While necessary, such work can lead to boredom, less attentiveness, and more mistakes for human workers. Humanoids, however, are designed to perform the same task repeatedly. They can perform the same action countless times without their minds becoming distracted, ensuring the quality remains consistent, thus allowing human workers to focus on more interesting or mentally challenging roles. This could include repetitive assembly line duties, sorting items in a storage facility, or basic data entry (Owen-Hill, 2022). By taking on these undesirable jobs, humanoids can make the workplace safer overall and make work more pleasing for human staff, allowing them to focus on more involved, creative, or managerial duties.

A significant advantage of humanoids is their ability to integrate seamlessly into existing human workspaces. Unlike fixed factory robots that require dedicated areas or a complete overhaul of a factory floor, humanoids are designed to resemble the shape of a person. This means they can often operate within human-friendly settings without requiring significant changes to the building. They can walk through standard doorways, go up and down stairs, move through pathways, and use tables or equipment designed for people. This ability to adjust is a big way to save money. Companies do not need to spend a lot to rebuild their workspace to accommodate new machines. A humanoid can be integrated into an existing setup and begin working with tools and systems that are already in place. This makes bringing in automation straightforward and trouble-free, allowing businesses to gradually add robotic assistance where it is most needed without having to overhaul their entire system.

Finally, humanoids can help fill gaps in the work team. In many places, certain industries face ongoing shortages of available people. This can occur due to an aging population, a lack of interest in certain types of jobs, or sudden increases in demand. Humanoid robots can help fill these gaps, providing a steady supply of workers when human staff are scarce. This is particularly important in areas such as care for older people, where the need for assistance is growing rapidly, or in factory sectors that struggle to find staff for repetitive tasks. By taking on some of the work, humanoids can help keep service levels and production going, making sure that important functions continue even when there are not enough human workers. They cannot replace all human jobs, but they can support and enhance the capabilities of the existing human work team.

The Difficulties: Obstacles and Drawbacks

Despite the clear benefits, integrating humanoids into the work team poses considerable challenges. These limits go from practical and technical problems to complex social and moral questions.

One of the biggest hurdles to using them everywhere is the high initial cost of buying and setting up humanoid robots (Yousefi-Koma et al., 2021). These machines are complex, involving many advanced parts, intricate moving sections, sophisticated sensors, and powerful computer systems. The extensive study and creation that go into making them is huge, and this is reflected in their cost. For many businesses, especially smaller ones, the initial cost can be too high, making it unclear whether they will recoup their investment with the robot (Tong et al., 2024). Beyond the purchase price, there are also ongoing costs. Humanoids require energy to function, often consuming a significant amount of power for their motors and computers. They also need regular maintenance, software updates, and occasional repairs. These continuing expenses contribute to the total cost of owning and operating them, which can make them less financially appealing compared to human labor in some situations. Furthermore, while humanoids are designed to be adaptable, making them truly useful in various work settings requires complex computer instructions and ongoing maintenance (Yousefi-Koma et al., 2021). Teaching a robot to perform a task that seems simple to a person, such as picking up a specific item from a cluttered shelf or recognizing a slight difference in a spoken direction, involves complex sets of rules and a substantial amount of practice information. This instruction often needs skilled engineers, which adds to the cost and complexity.

Additionally, these machines require regular maintenance and repair. Like any complex machine, parts can wear out, sensors can become misaligned, or software issues can occur. Finding and fixing these problems needs skilled people. This means that acquiring a humanoid robot is not the end of the investment; businesses also need to invest in developing the right skills to manage and care for them, which can be a significant effort.

While humanoids have made significant strides in grasping and manipulating objects, their dexterity remains limited compared to that of human hands. The human hand, with its complex network of bones, muscles, and nerves, can do exceedingly fine and sensitive movements. Human hands can feel how an object is made, judge its weight, and use just the right amount of force to pick up something delicate without breaking it. Robot hands, even the most advanced ones, struggle to copy this level of sensitivity and ability to adjust. They might be able to pick up many different objects, but handling very delicate items, doing extremely fine work, or working with soft, pliant materials (cloth or food) remains a significant problem. This limits what they can do in tasks that require a lot of tactile information or very gentle handling, areas where human skill is still superior.

Bringing powerful, moving machines into places shared with people naturally raises safety concerns. While designers incorporate safety features, the possibility of accidents, problems, or unintended actions cannot be eliminated. A robot might bump into a person, drop an item, or move in a way that is not expected, causing harm or damage. The "uncanny valley" (Section 3.3) also plays a part here. If a robot's movements or facial appearances are just a little bit off, it can make people uneasy, making them less comfortable working near the machine. This discomfort can impact the effectiveness of human-robot collaboration and the level of trust between them. Ensuring humanoids function as expected, safely, and in a manner that alleviates human concerns is a significant, ongoing challenge.

Beyond the technical and practical limitations, the presence of humanoids in the work team raises profound social and moral questions. The most common concern is job loss. If robots can do tasks currently done by people, what happens to those human workers? While supporters argue that robots free up people for more complex or creative roles, and that new jobs will emerge in robot upkeep and oversight, the change can be challenging for individuals whose jobs are taken by machines. Societies must consider how to support workers through these changes and ensure that the benefits of automation are shared widely.

There is also a concern that relying on robots for certain tasks might lead to a decline in human abilities over time. If machines handle all the repeated or precise work, will future generations lose the ability to do those tasks themselves? Another area of thought concerns human worth and the meaning of work. Some argue that integrating human-like robots into certain jobs, particularly those that involve care or personal interaction, could diminish the value of human work or alter the nature of human connections (Gary, 2021). If a robot provides companionship to an older person, does it make human connection less needed? What does it mean for human worth if people are replaced by machines in jobs that need emotion or good judgment? These are not simple questions and touch on deeply held societal beliefs.

Additionally, at this time, there is a lack of clear rules or a clear understanding of who is responsible for robots. Governments are still figuring out how to deal with these new machines. If a humanoid robot causes harm or makes a mistake, who is to blame? Is it the maker of the robot? The programmer who wrote its computer code? The purchaser? The employee who turned it on? These questions are no longer just ideas; they are now a reality. As humanoids move into workplaces, public areas, and private homes, the need for clear answers about who is responsible is pressing.

In summary, while humanoid robots offer several advantages in terms of the work they can perform, their safety, and their ability to handle challenging tasks, their widespread adoption in the workforce is still hindered by significant technical, financial, and social issues. Dealing with these limits will require continued progress in robotics and AI, along with careful consideration of the broader societal implications. The following section will examine how humans and robots can collaborate, aiming for a future where their combined strengths create more productive and safer work environments.

4.3 Human-Robot Collaboration (Cobots)

In the past, robots in factories often worked behind fences, separated from human staff for safety. They performed their duties alone, and any interaction with people was kept to a minimum. However, as robots become increasingly capable and, in the case of humanoids, more human-like, a new way of working is emerging human-robot collaboration, often involving machines known as "cobots," a shortened term for collaborative robots (Borboni et al., 2023). This approach moves beyond simple automation to a partnership where humans and robots work side by side, combining their unique strengths to achieve shared goals.

This section will explore the core idea behind human-robot collaboration, what makes a robot a "cobot," and the various ways people and humanoids can work together in shared environments.

The Idea Behind Working Together

The move toward human-robot collaboration comes from recognizing that neither humans nor robots are perfect on their own. Humans possess adaptability, common sense, problem-solving skills, and fine motor control that robots struggle to match. Humans can handle unexpected situations, make quick judgments based on little information, and perform delicate tasks that require a subtle touch. Robots offer strength, endurance, precision, and the ability to perform repetitive tasks without fatigue or error.

When these two sets of abilities are brought together, the outcome can be much greater than either could achieve alone. The idea is to have robots take on the tedious, heavy, or hazardous aspects of a job, freeing human workers to focus on the more complex, creative, or sensitive tasks. This is not about robots replacing people, but about robots assisting people in performing their jobs more effectively, safely, and efficiently.

For a robot to work closely with humans, it needs to be designed with safety at its core. Traditional robots are fast and powerful, which makes them a risk if a person steps into their work area. Cobots, on the other hand, are built to respond to human presence. They often have sensors that detect when a person is near or touches them, causing them to slow down or stop immediately. Their movements are often softer, and their parts are sometimes designed to reduce the force of any accidental contact. This focus on immediate safety response is what truly defines a collaborative robot.

Different Ways Humans and Robots Can Work Together

Human-robot collaboration is not a single entity; it can take many forms, depending on the task and the level of interaction required between the human and the machine. One simple form is co-existence. In this setup, humans and robots work in the same general area but perform separate tasks. They do not directly interact on the same piece of work at the same moment. For example, a humanoid might be moving supplies from one side of a factory floor to another, while human workers are assembling products at their workstations. They share the same space but operate independently of each other. The robot needs to be aware of its surroundings and avoid people, but it is not lending a hand to the same item.

A more direct form is sequential cooperation. Here, humans and robots perform different parts of the same task sequentially. A human might finish one step, then pass the item to a robot for the next step, or vice versa. Imagine an assembly line where a person places a small part, and then a humanoid robot comes along to tighten a series of screws precisely. There is a handoff of the work item, but they are not working on it simultaneously. The humanoid might use its vision systems to confirm the human's part is correct before it begins its work.

Then there is shared workspace collaboration. This is where humans and robots work on the same task simultaneously, within the same physical area. This is a closer form of teamwork. For instance, a human worker might be holding a large, heavy panel in place, while a humanoid robot helps by steadying it or putting in the first few fasteners. Here, the robot's strength and steadiness complement the human's ability to position and make quick decisions. The robot must be highly responsive and predictable to its human partner, allowing for easy, natural movement without fear of sudden or forceful actions. Its sensors constantly monitor the human's position, ensuring that the robot adjusts its speed or path to maintain safety.

A higher level of working together involves responsive collaboration. In this setting, the robot actively adjusts its actions based on the human's needs or current state. Imagine a humanoid in a medical setting, assisting a nurse. The robot might anticipate what tools the nurse needs next or move out of the way when the nurse approaches, without being directly told. This requires the robot to understand human intentions or predict what will happen next, often using its sensors to read human body language or gestures. It learns from observing the human worker, making the collaboration smoother and more effective over time.

Why Humanoids are Good for Working Together

Humanoid robots are particularly well-suited for collaboration for several reasons, mainly due to their human-like form and expanding abilities. First, their shape facilitates natural interaction. Because they have arms, hands, and a body that is shaped like ours, it's easier for humans to understand humanoid movements and anticipate their next action. Humanoids can pick up and use the same tools as people, share a workbench, or hand over an item in a way that feels familiar and comfortable. This reduces the time required for human workers to become accustomed to working with a machine. A humanoid can reach into tight spots, grasp objects of various shapes, and operate control panels designed for human hands.

Second, their growing dexterity is a big plus. As humanoid hands become more skilled, they can handle a wider array of objects and perform more complex manipulations. This means they can take on more parts of a task that previously required a human touch, making the collaboration more fruitful. They can pick up delicate components, put things together with care, or make fine adjustments.

Third, their sensing abilities are crucial for safety and smooth interaction. With advanced cameras, depth sensors, and touch sensors, humanoids can perceive their surroundings and recognize human presence, movements, and sometimes even gestures. This allows them to react quickly, stopping or slowing down if a human gets too close, or adjusting their actions to avoid collisions. This constant awareness is what makes close-quarters collaboration possible and safe.

Lastly, the ability of humanoids to learn and adapt over time makes them better partners. Through machine learning, they can observe human workers, learn preferred methods, and even adjust their movements to match a human's pace. This makes the partnership more fluid and helps the robot become a more intuitive helper, rather than just a programmed tool. They can learn from repeated interactions, gradually improving their collaborative actions without needing constant re-programming.

While the idea of human-robot collaboration is promising, its full realization still faces hurdles. Creating robots that are genuinely intuitive partners, capable of understanding complex human instructions or anticipating needs without explicit commands, remains a significant area of research. Communication between humans and machines needs to become more fluid, perhaps through natural speech, gestures, or even shared visual cues.

However, the move towards human-robot collaboration marks a significant step in how we perceive robots in the workplace. It shifts the discussion from robots replacing humans to robots augmenting human abilities. This hints at a future where machines do not challenge our strength, intelligence, and adaptability, but are enhanced by them. By working side by side, humans and humanoids can create work environments that are safer, more productive, and even more fulfilling for everyone involved. This partnership points to a future where machines are not just tools at a distance, but active, helpful colleagues.

4.4 Ethical and Social Implications of Humanoids in Daily Life

Having discussed the practical applications of humanoids in the workplace and the nuances of human-robot collaboration, we now turn our attention to their broader entry into daily life beyond the professional sphere. As human-like machines become more capable and accessible, their presence in our homes, schools, and public spaces brings forth a unique set of ethical and social questions. These are concerns that touch on our fundamental values, our sense of self, and the very nature of human connection. The challenge here is not just managing a new technology, but understanding how it might subtly, yet profoundly, reshape the fabric of our existence.

Private Spaces, Public Data: The Question of Personal Information

The most intimate spaces in our lives are our homes. When a humanoid robot enters this domain, it often comes equipped with an array of sensors: cameras that can see every movement, microphones that can hear every conversation, and touch sensors that register physical contact. These machines are designed to perceive their environment to be helpful, but, in doing so, they become constant gatherers of personal information. This raises immediate and pressing concerns about privacy and surveillance.

Imagine a humanoid assistant in your living room, capable of recognizing family members, understanding spoken instructions, and observing daily routines. While this functionality is intended to offer assistance, it also means the robot is constantly collecting data about your life: your habits, your conversations, your personal space (Eleshin et al., 2024). Who owns this vast amount of information? How is it stored? Is it protected from unauthorized access? The risk of this data being misused, whether by commercial entities for targeted advertising, by malicious actors for intrusion, or by other parties for surveillance, is a serious consideration. Even if the data is anonymized, patterns of behavior can often be deciphered (van den Hoven van Genderen, 2017). The mere presence of a device that is always listening or watching can alter behavior, leading to an uncomfortable situation where individuals feel constantly observed, even in their most private moments. Society must establish clear rules regarding data collection, usage, and retention for domestic robots, ensuring individuals maintain control over their personal information and the sanctity of their private spaces.

The Human Heart and the Machine: Psychological and Emotional Effects

The most complex set of questions revolves around the psychological and emotional impact of humanoids on our personal lives. When machines begin to mimic human interaction and appearance, they can elicit responses that are far more profound than those triggered by a simple appliance.

One significant concern is dependency. As humanoids become more capable of performing domestic tasks, providing companionship, or even assisting with personal care, individuals may become overly reliant on them. What happens if the robot malfunctions, breaks down, or is removed? Will individuals, particularly those who are vulnerable (such as the elderly or the very young), struggle to cope without their robotic assistant? This dependency extends beyond practical tasks to emotional support. If a person confides in a robot or uses it as their primary source of interaction, what happens to their ability to form and maintain relationships with human assistants?

This leads directly to the question of loneliness and substitution. Proponents argue that humanoids can alleviate loneliness, especially for isolated individuals or those in care settings. A robot companion that engages in conversation, plays games, or offers reminders might provide a sense of presence. However, critics ask whether this is a genuine solution or merely a temporary comfort that ultimately replaces meaningful human interaction. The illusion of companionship, where a non-sentient machine simulates empathy, raises ethical dilemmas. Is it right to permit individuals to form deep emotional attachments to something that cannot truly understand or reciprocate feelings? What are the long-term psychological effects of a relationship built on simulation rather than an authentic connection?

For children, the presence of humanoids in educational or play settings presents unique considerations. While robots can be engaging teachers or playmates, questions arise about their impact on human development. Will reliance on robotic interaction affect a child's ability to develop complex social skills, empathy, or creative problem-solving? Will the simulated emotions of a robot distort a child's understanding of genuine human feelings? These are uncharted waters, requiring careful research and thoughtful guidelines to ensure that humanoid interaction complements, rather than hinders, healthy human growth. The authenticity of relationships, whether for adults or children, becomes a central moral point when machines begin to mimic the very essence of human interaction.

What it Means to be Human

The increasing realism of humanoids also forces us to confront fundamental questions about identity and human dignity. If a machine looks, moves, and even converses like a person, how do we treat it? Is it acceptable to treat something that appears human as purely a tool, something that can be bought, sold, or decommissioned at will? The "objectification" of human-like machines raises discomfort for some, as it blurs the line between person and property.

Furthermore, the very existence of humanoids can affect our self-perception and what it means to be human. When machines can perform tasks once considered uniquely human (creative endeavors, emotional expression, complex problem-solving), how does that impact our sense of value and purpose? While Section 4.3 focused on job roles, this goes deeper into our intrinsic worth and identity in a world where machines are becoming increasingly capable of mimicking us.

A particularly sensitive area involves the replication of deceased kin or famous individuals. While we briefly touched on this in Section 3.3, its ethical implications for personal use are profound. The technology to create a humanoid likeness, complete with voice and mannerisms of someone who has passed away, raises questions about consent (did the deceased person agree to this?), the grieving process (does it help or hinder genuine acceptance of loss?), and the commercialization of personal identity. Who owns a person's likeness after they are gone? The creation of a "digital ghost" raises profound philosophical and emotional issues that society is just beginning to grapple with.

Fairness, Access, and the Digital Divide

The integration of humanoids into daily life also raises questions about accessibility and equity. Who will benefit from these advanced technologies? If personal humanoid assistants or companions are expensive, will they only be available to the wealthy, thereby exacerbating existing social and economic divides? Will they create a new class of technological "haves" and "have-nots," where access to advanced assistance or companionship is determined by economic status?

Furthermore, there is the ongoing issue of cultural bias and representation in robot design and the data used to train them. As discussed previously, if humanoids reflect only certain cultural norms or are trained on biased datasets, they may not interact effectively or fairly with diverse populations. This could lead to unequal access to their benefits or even perpetuate societal biases through technological means. Ensuring fairness in design and distribution, and making sure these technologies are developed and deployed with global diversity in mind, is an ethical imperative.

The introduction of humanoid robots into our daily lives presents a fascinating yet daunting challenge. They promise convenience, assistance, and perhaps even a form of companionship, but they also force us to confront uncomfortable questions about privacy, psychological well-being, human identity, and societal fairness.

As these machines become more integrated into our homes, schools, and personal routines, a balanced and proactive approach is essential. This requires:

- Robust ethical frameworks and regulations are needed to address data privacy, accountability, and the responsible design of AI that mimics emotion.
- Ongoing research into the long-term psychological and social effects of human-robot interaction (HRI), particularly for vulnerable populations.
- Public discourse and education to ensure informed societal decisions about the role of humanoids in our lives, moving beyond hype or fear to a realistic understanding.
- Emphasis on human-centered design that prioritizes human dignity, autonomy, and genuine well-being over mere technological capability.

Humanoids in daily life are not just about what they can do for us, but also about what they reveal about us. How we choose to integrate them will reflect our values and shape the kind of future we build for ourselves alongside our increasingly human-like creations. This journey requires not just engineering brilliance but also profound societal wisdom.

Bibliography

Barakeh, Z. A., Alkork, S., Karar, A. S., Said, S., & Beyrouthy, T. (2019, April). Pepper humanoid robot as a service robot: A customer approach. *2019 3rd International Conference on Bio-Engineering for Smart Technologies (BioSMART)*. https://doi.org/10.1109/biosmart.2019.8734250

Borboni, A., Reddy, K. V. V., Elamvazuthi, I., Al-Quraishi, M. S., Natarajan, E., & Azhar Ali, S. S. (2023). The expanding role of artificial intelligence in collaborative robots for industrial applications: A systematic review of recent works. *Machines*, *11*(1), 111. https://doi.org/10.3390/machines1101011

Brown, L., & Howard, A. M. (2013). Engaging children in math education using a socially interactive humanoid robot. *2013 13th IEEE-RAS International*

Conference on Humanoid Robots (Humanoids), 183–188. https://doi.org/10.1109/humanoids.2013.7029974

Eleshin, F., Iradukunda, P., Ruberamitwe, D., & Ishimwe, E. (2024). Your robot might be inadvertently or deliberately spying on you: A critical analysis of privacy practices in the robotics industry. *Proceedings of the 10th International Conference on Information Systems Security and Privacy*, 219–225. https://doi.org/10.5220/0012422200003648

Garabini, M., Caporale, D., Tincani, V., Palleschi, A., Gabellieri, C., Gugliotta, M., Settimi, A., Catalano, M. G., Grioli, G., & Pallottino, L. (2021). WRAPP-up: A dual-arm robot for intralogistics. *IEEE Robotics & Automation Magazine*, *28*(3), 50–66. https://doi.org/10.1109/mra.2020.3015899

Gary, M. E. (2021). Care Robots, Crises of Capitalism, and the Limits of Human Caring. *International Journal of Feminist Approaches to Bioethics*, *14*(1), 19–48. https://doi.org/10.3138/ijfab-2020-07-28

Hentout, A., Aouache, M., Maoudj, A., & Akli, I. (2019). Human–robot interaction in industrial collaborative robotics: A literature review of the decade 2008–2017. *Advanced Robotics*, *33*(15–16), 764–799. https://doi.org/10.1080/01691864.2019.1636714

Mukherjee, S., Baral, M. M., Pal, S. K., Chittipaka, V., Roy, R., & Alam, K. (2022). Humanoid robot in healthcare: A Systematic Review and Future Research Directions. *2022 International Conference on Machine Learning, Big Data, Cloud and Parallel Computing (COM-IT-CON)*, 822–826. https://doi.org/10.1109/com-it-con54601.2022.9850577

Owen-Hill, A. (2022). Five highly dangerous jobs that robots can do safely. *Smart Manufacturing*, 415–418. https://doi.org/10.1002/9781119846642.other8

Ozturkcan, S., & Merdin-Uygur, E. (2021). Humanoid service robots: The future of healthcare? *Journal of Information Technology Teaching Cases*, *12*(2), 163–169. https://doi.org/10.1177/20438869211003905

Patricio, R., & Mendes, A. (2020). Consumption patterns and the advent of automated guided vehicles, and the trends for automated guided vehicles. *Current Robotics Reports*, *1*(3), 145–149. https://doi.org/10.1007/s43154-020-00007-4

Rasmussen, M. K., Schneider-Kamp, A., Hyrup, T., & Godono, A. (2024). New colleague or gimmick hurdle? A user-centric scoping review of the barriers and facilitators of robots in hospitals. *PLOS Digital Health*, *3*(11), e0000660. https://doi.org/10.1371/journal.pdig.0000660

Royakkers, L., & van Est, R. (2015). *Just ordinary robots.* CRC Press. https://doi.org/10.1201/b18899

Tanioka, T., Smith, M. C., Osaka, K., & Zhao, Y. (2019). Framing the development of humanoid healthcare robots in caring science. *International Journal for Human Caring*, *23*(2), 112–120. https://doi.org/10.20467/1091-5710.23.2.112

Tong, Y., Liu, H., & Zhang, Z. (2024). Advancements in humanoid robots: A comprehensive review and future prospects. *IEEE/CAA Journal of Automatica Sinica*, *11*(2), 301–328. https://doi.org/10.1109/jas.2023.124140

van den Hoven van Genderen, R. (2017). Privacy and data protection in the age of pervasive technologies in AI and robotics. *European Data Protection Law Review*, *3*(3), 338–352. https://doi.org/10.21552/edpl/2017/3/8

Younis, H., Mohamed, A. S. A., Jamaludin, R., & Wahab, M. N. A. (2021). Survey of robotics in education, taxonomy, applications, and platforms during COVID-9. *Computers, Materials & Continua*, *67*(1), 687–707. https://doi.org/10.32604/cmc.2021.013746

Yousefi-Koma, A., Maleki, B., Maleki, H., Amani, A., Bazrafshani, M. A., Keshavarz, H., Iranmanesh, A., Yazdanpanah, A., Alai, H., Salehi, S., Ashkvari, M., Mousavi, M., & Ashtiani, M. S. (2021). SURENAIV: Towards A cost-effective full-size humanoid robot for real-world scenarios. *2020 IEEE-RAS 20th International Conference on Humanoid Robots (Humanoids)*, 142–148. https://doi.org/10.1109/humanoids47582.2021.9555686

Zhong, L., & Verma, R. (2022). Rise of humanoid robots in hospitality services. In *The service robot revolution* (pp. 331–345). Edward Elgar Publishing. https://doi.org/10.4337/9781800375659.00034

CHAPTER 5

Autonomy and Decision-Making

In Chapter 2, we explored the pivotal shift from rigid automation to the era of intelligent machines, introducing AI and Machine Learning, particularly Reinforcement Learning, as the core technologies enabling robots to learn. We learned that humanoids could begin to "see" and "learn to walk" through these advancements. Now, we delve deeper into precisely how these learned capabilities are orchestrated to form actual, dynamic decisions, and the intricate architectural and practical challenges that arise when a humanoid operates autonomously.

5.1 The Humanoid Mind: How Decisions are Made

For a humanoid robot to truly operate independently in complex, unpredictable human environments, it must undertake far more than merely executing pre-programmed motions or mimicking behaviors it has learned. It needs to assess its surroundings actively, interpret the meaning of what it perceives, formulate coherent plans, choose optimal actions, and adapt its approach in real time. This sophisticated interplay of sensing, understanding, and acting forms the continuous cycle of humanoid decision-making, often conceptualized as a perception-cognition-action loop.

A humanoid's ability to make intelligent choices unfolds through a constant, integrated process where each stage seamlessly feeds into the next, ensuring fluid and responsive behavior. The foundation of this process lies in sophisticated

DOI: 10.1201/9788743810704-7

perception. While Chapter 2 touched on computer vision and natural language processing as breakthroughs, it is crucial to recognize the intense, continuous operation of these systems as they supply the essential input for every decision. Perception for a humanoid is not simply about identifying a few objects; it is about constructing a rich, ever-evolving internal model of the world in real-time. Humanoids are equipped with a diverse array of sensors for this purpose. Cameras, both two-dimensional and three-dimensional, provide visual data for tasks such as object recognition, facial identification, gesture interpretation, and depth sensing (Gao et al., 2021). Advanced Machine Learning models, particularly Convolutional Neural Networks, tirelessly process this raw pixel data, transforming it into meaningful semantic information. This enables the robot to distinguish, for instance, a cup from a book, to identify a specific person and track their movement, or to recognize a potential obstacle (Tong et al., 2024).

Beyond visual input, LiDAR and radar systems generate precise three-dimensional maps of the environment, which are indispensable for accurate navigation, reliable obstacle avoidance, and a nuanced understanding of spatial relationships. These sensors provide robust data even when lighting conditions are challenging. Microphones capture auditory cues, feeding into Natural Language Processing models that allow the robot not only to comprehend spoken commands and questions but also to infer human emotional states from the tone of voice (Spatola et al., 2021). Tactile and force sensors, strategically placed in hands, feet, and throughout the robot's body, provide crucial feedback on contact, pressure, and grip force. This sensory input is critical for delicate manipulation tasks, such as handling fragile objects, and for maintaining the robot's balance during locomotion. Finally, IMUs, furnish data on the robot's orientation, angular velocity, and acceleration, which are vital for maintaining its equilibrium, tracking its body movements through proprioception, and understanding the external forces acting upon it (Ahmad et al., 2013).

The inherent challenge at this initial stage is the intricate task of fusing this vast, diverse, and often noisy or incomplete sensor data into a coherent, real-time "understanding" of the surrounding world. Machine Learning algorithms work continuously to filter out irrelevant information, fill in any data gaps, and generate a comprehensive "state estimate." This estimate is a detailed picture of the robot's precise position, the exact configurations of its joints, and the dynamic status of its environment, including the location and predicted movements of humans and other objects.

Once the environment is sufficiently perceived and modeled, the humanoid's cognition and planning systems engage to process this wealth of information and formulate a logical course of action. This stage extends beyond mere recognition, delving into the realms of reasoning, prediction, and strategic planning. Building upon the raw perceptual data, the robot constantly refines its internal map of the world through techniques like SLAM. This is far from a static blueprint; it is a dynamic model that actively tracks moving objects, identifies navigable and non-navigable areas, and recognizes specific environmental features such as doors, chairs, or tools (Das, 2020). Critically, it also predicts future states of the environment, for example, where a moving human might be in the next few seconds, thereby enabling proactive rather than purely reactive decisions.

When a high-level command is received, such as “prepare coffee,” the robot’s AI must decompose this overarching objective into a logical sequence of achievable sub-goals, which could include “get mug,” “insert coffee pod,” and “press button.” This decomposition often relies on learned task models, where the robot has either observed or been trained on how to achieve similar objectives in the past. With a goal or sub-goal defined, the robot must then meticulously plan its physical movements. This involves complex algorithms that calculate collision-free paths for its entire body, known as whole-body planning, and fine-tuned trajectories for its limbs, particularly for dexterous manipulation. Reinforcement Learning plays a pivotal role here, as humanoids can learn highly optimized and natural-looking movements that are remarkably robust to unexpected perturbations. For instance, a robot-like Boston Dynamics’ Atlas learning parkour is not simply following a pre-set choreography; it is constantly calculating and adjusting its trajectory and balance in milliseconds to execute incredibly dynamic maneuvers.

A key hallmark of advanced decision-making is the capacity for anticipation and prediction. By learning from vast datasets of human behavior and environmental dynamics, humanoid AI can predict likely future events. If a human reaches for an object, the robot can anticipate their intention to pick it up and adjust its action or path accordingly. This predictive capability significantly enhances seamless human-robot collaboration and is fundamental for safety in shared environments.

Finally, the formulated plan is translated into precise commands for the robot’s actuators, its motors, joints, and grippers. As the robot moves and interacts, its sensors continuously feed new data back into the perception system, and the entire loop of perception-cognition-action restarts. This real-time feedback mechanism allows the robot to make immediate, on-the-fly adjustments if anything unexpected occurs, such as its grip slipping, an object moving, or an unforeseen obstacle appearing in its path. This continuous feedback loop is the very essence of true adaptability.

The Architecture of Autonomy in Humanoid Robots: Decision Orchestration

Humanoids often use advanced control architectures to handle the huge complexity of this decision-making pipeline. One common method is hierarchical control, which breaks down decision-making into different levels of responsibility and abstraction (Pateria et al., 2021). The strategic layer, which is the uppermost level, is where people think about long-term goals and make plans that are not specific. This includes AI parts that think about how to organize sequences of difficult jobs, choose overarching tactics, or figure out what people really mean. The tactical layer, which is in the middle, is in charge of turning these big goals into smaller, more manageable, tasks and activities (Pateria et al., 2021). A humanoid navigating a room could need to make a series of footsteps or a series of whole-body poses to get to a certain spot. The reactive or operational layer, which is the lowest level, controls motors and sensors in real time to make precise movements, keep balance, and quickly respond to unexpected occurrences, such changing the force of the legs to make up for sudden changes in terrain. This lowest level often needs quick, reactive controls, like those learnt by reinforcement learning. Many advanced humanoids successfully combine high-level planning with strong low-level controllers that can make quick changes to keep stability and carry out dynamic maneuvers.

There are two main types of decision-making systems: reactive and deliberative (Palanca et al., 2023). Reactive systems react to sensor inputs right away, almost like instinct, and they don't usually develop detailed internal world models or do complicated planning. Like a person reflexively withdrawing his hand away from a hot surface, they are quick and strong in dynamic settings. A lot of the basic locomotion and balancing systems of humanoids depend on learnt reactive policies, which are often based on reinforcement learning (Gu et al., 2022).

Deliberative systems, on the other hand, require building and thinking about a complex internal model of the environment and carefully planning a series of activities to reach certain goals. These systems usually take longer to respond and use more processing power, but they can solve very complicated problems and plan for the long term. Planning the best way to get through a building with more than one room or following a precise assembly process for a product with more than one part are two examples (Al-Ghanimi & Al-Moadhen, 2022). Most effective humanoids use hybrid architectures that combine the quickness of reactive control with the ability to think ahead and solve problems that comes with deliberate planning. A robot might use a deliberative planner to figure out how to get across a factory floor, but it could also switch to reactive control mode right once to avoid an unexpected barrier or get back on its feet if pushed.

Intrinsic Challenges in Autonomous Decision-Making

Despite the remarkable progress in humanoid robotics, enabling these machines to make brilliant, reliable, and safe decisions presents significant ongoing challenges that remain at the forefront of research.

One formidable challenge is generalization and robustness. A robot might achieve seemingly perfect performance within its controlled training environment yet struggle considerably when introduced to a slightly different real-world scenario (Gu et al., 2023). This could manifest as difficulty adapting to novel lighting conditions, interacting with a slightly altered version of a familiar object, or navigating a previously unseen clutter arrangement. Achieving robust generalization – the ability for a robot to perform consistently well in unforeseen variations of known tasks or to apply its learning effectively in entirely novel situations – remains an immense hurdle. The inherent chaos and infinite variability of the real world far exceed the scope of any finite training dataset or simulation.

Another critical issue is often referred to as the "black box" problem. Many of the most powerful AI models, intense neural networks foundational to perception and reinforcement learning, operate as opaque black boxes. It becomes challenging for human engineers to truly understand why the AI made a particular decision or arrived at a specific solution (Hassija et al., 2023). This inherent lack of interpretability can severely undermine human trust in autonomous systems, complicate debugging efforts when errors inevitably occur, and pose profound challenges for assigning accountability in critical applications. When a robot makes a significant mistake, pinpointing the exact chain of reasoning or the specific learned pattern that led to that error can be an elusive and frustrating task.

Furthermore, the quality of data, bias, and completeness directly impact the intelligence of any Machine Learning-driven decision-making system. If the training data is inherently biased – for instance, only showing objects from specific angles, or disproportionately representing certain human demographics – the robot's subsequent decisions will inevitably inherit and perpetuate those biases, potentially leading to unfair, discriminatory, or incorrect actions (Osasona et al., 2024). Compounding this, it is fundamentally impossible to train an AI on all conceivable real-world scenarios, leaving inherent gaps in its learned decision-making capabilities that could manifest in unexpected ways.

The computational load and demand for real-time processing also pose a continuous engineering challenge. Processing vast streams of raw sensor data, running complex AI models for perception and planning, and simultaneously coordinating the movements of dozens of motors and actuators, all in real-time, requires immense computational power. Achieving this efficiently within the physical constraints of the robot, factoring in power consumption, heat dissipation, and weight, is a constant balancing act. To ensure both safety and fluidity, many critical decisions must be made within milliseconds.

Safety and predictability are paramount concerns, primarily when humanoids operate in close proximity to humans (Duburcq et al., 2022). When a robot's behavior is learned rather than explicitly programmed, it can exhibit emergent behaviors that were not specifically intended or even foreseen by its creators. While such emergent properties can sometimes lead to innovative solutions, they also introduce an element of unpredictability. Guaranteeing that autonomous decisions are consistently safe, especially in complex and dynamic human-shared environments, necessitates rigorous testing, the integration of robust fail-safe mechanisms, and sophisticated anomaly detection systems. The profound question of how to ensure a robot will always make the "right" decision in an ethically ambiguous or potentially dangerous situation remains a central philosophical and engineering dilemma.

Finally, the capacity for adaptation to true novelty continues to be a frontier. While humanoids can adapt remarkably well to variations within familiar domains, such as navigating different types of rough terrain, truly adapting to situations that are entirely new and fall outside the scope of any prior training or simulation remains a profound challenge. This is where more advanced forms of reasoning, abstraction, and common-sense knowledge, which are still nascent in current AI systems, become critically crucial for unlocking higher levels of autonomous intelligence.

To sum up, a humanoid's "mind" is a dynamic combination of improved perception, complex cognitive processing, and adaptive action, all made possible by advanced AI techniques. Machine Learning and Reinforcement Learning have opened up new levels of autonomy, allowing machines to see, learn, and act with more and more independence. However, the goal of creating humanoids that can make intelligent, strong, and universally applicable decisions is still a long way off. A major factor that affects these problems and a key architectural issue is where this intricate "thinking" happens most of the time. This is a difference between processing power that is directly on the robot and computation that is done on remote servers. This brings us to the important topic of Cloud AI vs. Edge AI.

5.2 Cloud AI vs. Edge AI: Where Humanoids Do Their Thinking

The sophisticated decision-making processes in humanoid robots – from interpreting complex sensor data to planning intricate actions – demand significant computational power. A crucial architectural decision lies in where this "thinking" takes place: onboard the robot itself, known as Edge AI, or remotely in powerful data centers, referred to as Cloud AI. While many imagine a robot's intelligence as a self-contained brain operating in real time, the reality is often more distributed. Much of the processing may occur off-site in cloud infrastructure, or directly on the robot when immediate, localized responses are needed. Each approach carries its strengths and trade-offs, influencing a robot's autonomy, responsiveness, cost, and security.

The Cloud: Big Brain, Far Away

Cloud-based AI is exactly as it sounds. Instead of storing and processing data locally, the robot connects to remote servers: massive computers packed with high-end processors, vast memory banks, and access to huge libraries of data. These servers can run complex algorithms that would overwhelm a robot's small onboard systems (Saravanan, 2017).

With this architecture, a humanoid robot's sensors capture data. This raw or lightly pre-processed information is then transmitted over a network (often Wi-Fi or 5G) to the cloud for heavy-duty AI processing, model inference, and complex planning. The results or commands are then sent back to the robot for execution.

The most significant advantage of Cloud AI is unparalleled computational power and scalability. Cloud servers possess virtually limitless processing capabilities, allowing humanoids to leverage extensive and complex AI models, such as sophisticated LLMs for nuanced natural language understanding or massive deep learning networks for highly accurate object recognition across millions of categories (Saravanan, 2017). This enables a level of intelligence that would be impossible with onboard hardware alone. Cloud platforms also offer elastic scalability, meaning computing resources can be increased or decreased as needed, which is cost-effective for development and managing fluctuating workloads across a fleet of robots.

Cloud AI also facilitates collective learning and shared intelligence. Data collected by numerous robots can be aggregated and analyzed in the cloud to train even more powerful, generalized AI models (Afrin et al., 2021). This "fleet learning" allows individual robots to benefit from the experiences of others, accelerating their capabilities. Software updates, new behaviors, and improved AI models can be deployed simultaneously to an entire fleet from a centralized cloud platform, ensuring consistency and rapid iteration. Furthermore, cloud connectivity enables seamless human-robot collaboration through remote monitoring, teleoperation support, and the integration of robots with other enterprise systems (Liu et al., 2020).

However, Cloud AI introduces critical challenges for humanoids. The most pressing concern is latency. The time it takes for sensor data to travel to the cloud, be processed, and for commands to return to the robot, known as network latency, can be a significant impediment for real-time applications. In dynamic physical interactions, even a delay of tens or hundreds of milliseconds can make a robot unsafe or ineffective. Imagine a robot that has to decide whether to stop because someone walked in front of it; if it needs to send that decision to the cloud, wait for a reply, then act, it might already be too late. Connectivity issues are also a significant drawback; without a stable, high-bandwidth internet connection, a cloud-dependent humanoid becomes significantly hampered or entirely inoperable, losing its "brain."

Data security and privacy risks are heightened with Cloud AI, as sensitive operational data must constantly traverse public or private networks. This increases the attack surface for cyber threats and necessitates robust encryption and stringent access controls. Finally, operational costs can accrue significantly due to continuous data transmission and the "pay-as-you-go" nature of cloud services, especially for robots generating vast amounts of raw data (Zahoor, 2023).

Cloud AI works best for tasks that are thoughtful rather than urgent. Imagine a robot organizing its schedule for the week, or retrieving information from a vast recipe database to help with meal prep. These are situations where timing isn't critical. A few extra seconds won't harm. But when a robot needs to make an immediate decision – like avoiding a falling object or reacting to a person suddenly stepping into its path – that delay becomes a problem. Sending data to a remote server and waiting for a reply can take too long. By the time the response arrives, the moment may have already passed. That's where Edge AI excels.

Edge AI: Thinking on the Spot

Edge refers to the idea of computing happening right at the "edge" of the network – close to where the action is. In robotics, that means the robot itself handles the processing. The camera's feed directly into the robot's processor.

The algorithms run locally. The decision happens instantly. Edge AI is about speed. There's no waiting and no round-trip data delay. The robot sees, thinks, and acts all in one place (Surya & Santoso, 2024).

One of the most compelling advantages of Edge AI for humanoids is reduced latency. In situations demanding split-second reactions, such as maintaining balance on uneven terrain, avoiding a suddenly appearing obstacle, or catching a thrown object, even a few milliseconds of delay can lead to catastrophic failure (Surya & Santoso, 2024). By processing data directly on the device, there is no need to transmit information to a remote server and wait for a response, eliminating network round-trip delays. This allows for near-instantaneous decision-making and real-time responsiveness, crucial for dynamic physical interaction.

Furthermore, Edge AI enhances operational reliability. Humanoids often operate in environments where constant, high-bandwidth internet connectivity cannot be guaranteed (Wang et al., 2023). Imagine a robot exploring a disaster zone, assisting in a remote hospital, or performing tasks in a factory with unreliable Wi-Fi. With Edge AI, the robot remains fully functional and autonomous even if network access is intermittent or completely unavailable. This self-sufficiency makes it robust and dependable in diverse and challenging settings.

Data privacy and security also benefit significantly from Edge AI. When sensitive sensor data, such as high-resolution video streams from cameras or personal interaction data, is processed locally on the robot, it largely remains on the device (Kong et al., 2022). This minimizes the need to transmit potentially confidential or personally identifiable information over networks to external servers, thereby reducing exposure to breaches during transmission and simplifying compliance with strict data protection regulations like General Data Protection Regulation (GDPR). Additionally, Edge AI can lead to lower bandwidth usage and operational costs over time, as only crucial insights or processed data, rather than raw, voluminous sensor streams, might occasionally be sent to the cloud for logging or long-term analysis.

However, Edge AI is not without its drawbacks. The primary limitation stems from constrained processing power and memory (Kong et al., 2022). Humanoids have physical size, weight, and power consumption limits, which restrict the size and number of processors they can carry. This often means they cannot run the most extensive, most computationally intensive, AI models directly. Developers must optimize their machine learning models, frequently using techniques like model quantization or pruning, to make them lightweight enough to run efficiently on onboard hardware. This sometimes means sacrificing a degree of accuracy or complexity compared to what a cloud-based system could achieve.

Initial deployment costs can also be higher due to the need for specialized, powerful, and miniaturized hardware, and maintenance and updates can be more complex, requiring individual robot updates rather than centralized server updates.

The Hybrid Approach

In practice, most modern robots use both. Edge AI handles fast, real-time decisions. Cloud AI handles deep learning, software updates, and long-term memory.

Let's say a warehouse robot is sorting packages. It uses Edge AI to detect if something's in its path, correct its grip on a box, or balance while rolling. But at the end of the shift, it sends its performance data to the cloud. Engineers can review the data, make improvements, and update the robot's software overnight. That hybrid approach creates a feedback loop. Local intelligence makes the robot responsive. Cloud intelligence helps it get smarter over time.

This model also allows for fleet learning. If one robot encounters a problem and finds a solution, that solution can be uploaded to the cloud and distributed to every other unit. It's like each robot becomes a scout, and the cloud is the shared brain that educates everyone.

Amazon's warehouse robots are a real-world example. They use Edge AI to navigate and avoid collisions in real-time, but the overall traffic flow, routing strategies, and performance analytics are handled by the cloud. It's a division of labor that plays to the strengths of each system. This symbiotic relationship between Edge and Cloud allows humanoids to be both responsive and intelligent. It enables them to navigate unpredictable physical environments with real-time adaptation while simultaneously benefiting from the continuous learning and vast data resources that only cloud computing can provide (Afrin et al., 2021). The ongoing advancements in 5G technology, with its promise of lower latency and higher bandwidth, further blur the lines, enhancing the viability and fluidity of these hybrid approaches. The choice of where a humanoid's brain resides or, more accurately, how its distributed intelligence is orchestrated, is a fundamental design decision that shapes its capabilities, its operational environment, and its potential for truly autonomous and adaptable behavior.

As humanoid robots move out of labs and into everyday life, how they process information becomes one of the key design decisions. Some will be intelligent but slow. Others will be fast but forgetful. The best will find a way to be both. That's the quiet revolution happening behind the scenes: not in how humanoids look or move, but in where and how they think.

The journey of creating truly autonomous humanoids, capable of intelligent decision-making, leads us inevitably to the crucible of the real world. While the previous sections discussed the core mechanisms of their "thinking" and the strategic placement of their computational "brains," the ultimate test lies in their ability to perform real-time adaptation within complex environments. This is where the tidy algorithms of the laboratory collide with the messy, unpredictable, and often chaotic reality of human existence.

5.3 Challenges of Real-time Adaptation in Complex Environments

The locations that humanoids are being created to live in, such as homes, public places, hospitals, and disaster zones, are very different from the regulated and predictable conditions of a standard manufacturing floor (Ziglar et al., 2017). Think about the problems: a child could leave a toy in the corridor, a pet could suddenly cross a robot's path, lighting conditions may change unexpectedly, or a human could abruptly stop or change direction. Most of the time, surfaces aren't precisely flat, and soundscapes are typically a mix of sounds that don't fit together. As a result, humanoids must process with a continuous stream of new stimuli, which is very different from the repetitious routines of earlier industrial robots.

Adding to this is the fact that uncertainty and incompleteness are everywhere. No matter how advanced they are, a robot's sensors have limits. Objects can be partially hidden, sensor data can be unclear or noisy, and the robot can never have a full or complete view of its surroundings. The open-world dilemma is that you can't program or train a robot for every potential combination of events it might see in that kind of setting. There are so many options that you need to be able to make decisions quickly and easily. Also, figuring out complicated social cues, such figuring out what others mean beyond simple directives, reading subtle gestures, or adjusting to different social cues, makes real-time adaptation even harder. This unpredictability gives rise to several fundamental challenges for a humanoid's ability to adapt in real time.

A significant challenge is perceptual ambiguity and robustness. Humanoids must constantly process noisy, incomplete, or rapidly changing sensor data (Renganathan et al., 2020). For instance, robustly recognizing and tracking multiple objects in a cluttered, dynamic scene – like distinguishing a toy from a valuable item amidst a pile of laundry – is far more complex than identifying a single, isolated object. The robot must also contend with varying lighting conditions, reflections, smoke, dust, or even new angles from which objects are viewed, all of which can severely degrade perception and lead to misinterpretations. Its decision-making chain is only as strong as its perception.

Equally demanding is the task of real-time planning and replanning. The computational burden of instantly recalculating optimal trajectories or entire task sequences when the environment changes unexpectedly is immense. If a door is closed, a person steps into its planned path, or an object falls, the robot cannot halt and ponder for minutes. It must generate a new, collision-free path for its complex body and manipulators in milliseconds (Gentner et al., 2023). This necessitates a delicate balance between finding the mathematically optimal solution and developing a sufficiently good solution quickly enough to remain safe and effective. Furthermore, the robot must adapt its physical interaction on the fly, for instance, by adjusting its grip strength if an object proves heavier or more fragile than expected, or by dynamically shifting its weight to regain balance on a slippery surface.

The most profound test comes when confronting novelty and out-of-distribution (OOD) data. Machine Learning models, while powerful, inherently struggle when presented with data that falls outside their training distribution. How does a humanoid truly adapt to a genuinely new object, an unprecedented environmental configuration, or a scenario it has never encountered in its training data or simulations? The "brittleness" of AI when faced with true novelty can lead to unpredictable or nonsensical behavior. This highlights the persistent need for humanoids to possess a capability for continual, on-the-fly learning directly from their real-world experiences, rather than relying solely on pre-trained models.

The practicalities of physical interaction and dexterity in unconstrained settings also present ongoing challenges. Achieving human-level dexterity for general-purpose manipulation, where a humanoid can reliably grasp, lift, place, and use a diverse range of tools and objects with varying weights, textures, and fragilities, remains a significant hurdle. Precision in force control, ensuring compliant and safe interaction with both objects and humans, and adapting to unknown physical properties are all critical for seamless integration into dynamic environments.

Finally, navigating ethical and safety adaptation in real-time adds a layer of moral complexity. How does a humanoid make nuanced ethical decisions, such as prioritizing the protection of one person over another in a novel, multi-agent collision scenario, if it encounters an ethical dilemma for which it has no explicit programming or clear learned guidance? Adapting to implicit human trust levels and subtle social norms in real-time, for instance knowing when to offer help versus when to maintain distance, further complicates its adaptive decision-making process.

Strategies for Overcoming Adaptation Challenges

To overcome these formidable challenges, researchers and engineers are employing various sophisticated strategies. Improved sensor fusion is critical, combining and intelligently interpreting data from multiple sensor types – cameras, LiDAR, radar, and haptic feedback – to construct a more robust, comprehensive, and resilient world model. Sim-to-real transfer is another powerful approach, where humanoids are trained in highly realistic, physics-accurate simulations that expose them to millions of variations and edge cases, preparing them for the complexities of the real world before deployment. While limited by safety considerations, reinforcement learning with real-world feedback continues to refine learned behaviors on actual hardware. The use of hybrid AI architectures, as discussed in Section 5.2 allows humanoids to balance the immediacy of reactive reflexes with the foresight of deliberative planning, providing a robust framework for adaptation. Furthermore, in situations where complete autonomy is not yet achievable or deemed too risky, human-in-the-loop systems remain vital, allowing remote human operators to provide oversight, intervene, or take control in particularly challenging or ambiguous circumstances.

In summary, real-time adaptation in complex, unstructured environments represents the ultimate frontier and the most rigorous test of humanoid intelligence. While the progress in equipping these machines with sophisticated learning and decision-making capabilities has been rapid and astonishing, this domain remains a profound area of ongoing research. Overcoming these challenges pushes the very boundaries of AI, robotics, and HRI, paving the way for truly versatile and integrated humanoid companions and workers. However, as these robots become increasingly autonomous and adaptive, new questions inevitably arise concerning their security and ethical operation, topics we will explore in Chapter 6.

Bibliography

Afrin, M., Jin, J., Rahman, A., Rahman, A., Wan, J., & Hossain, E. (2021). Resource allocation and service provisioning in multi-agent cloud robotics: A comprehensive survey. *IEEE Communications Surveys & Tutorials*, *23*(2), 842–870. https://doi.org/10.1109/comst.2021.3061435

Ahmad, N., Ghazilla, R. A. R., Khairi, N. M., & Kasi, V. (2013). Reviews on various inertial measurement unit (IMU) sensor applications. *International Journal of Signal Processing Systems*, 256–262. https://doi.org/10.12720/ijsps.1.2.256-262

Al-Ghanimi, Z., & Al-Moadhen, A. A. (2022). Deliberative robotics behavior systems: A survey. *2022 International Conference on Data Science and Intelligent Computing (ICDSIC)*, 147–152. https://doi.org/10.1109/icdsic56987.2022.10076116

Das, P. (2020). Advancing simultaneous localization and mapping (SLAM) for robots in unstructured terrain. *International Journal for Multidisciplinary Research*, *2*(6). https://doi.org/10.36948/ijfmr.2020.v02i06.25432

Duburcq, A., Schramm, F., Boéris, G., Bredeche, N., & Chevaleyre, Y. (2022). Reactive stepping for humanoid robots using reinforcement learning: Application to standing push recovery on the exoskeleton Atalante. *2022 IEEE/RSJ International Conference on Intelligent Robots and Systems (IROS)*, 9302–9309. https://doi.org/10.1109/iros47612.2022.9982234

Gao, S., Dai, Y., & Nathan, A. (2021). Tactile and vision perception for intelligent humanoids. *Advanced Intelligent Systems*, *4*(2). https://doi.org/10.1002/aisy.202100074

Gentner, M., Zillenbiller, F., Kraft, A., & Steinbach, E. (2023). Dynamic multi-query motion planning with differential constraints and moving goals. *2023 IEEE/RSJ International Conference on Intelligent Robots and Systems (IROS)*, 205–212. https://doi.org/10.1109/iros55552.2023.10341670

Gu, J., Kirmani, S., Wohlhart, P., Lu, Y., Arenas, M. G., Rao, K., Yu, W., Fu, C., Gopalakrishnan, K., Xu, Z., Sundaresan, P., Xu, P., Su, H., Hausman, K., Finn, C., Vuong, Q., & Xiao, T. (2023, November 3). *RT-Trajectory: Robotic task generalization via hindsight trajectory sketches*. arXiv.Org. https://arxiv.org/abs/2311.01977

Gu, Z., Boyd, N., & Zhao, Y. (2022). Reactive locomotion decision-making and robust motion planning for real-time perturbation recovery. *2022 International Conference on Robotics and Automation (ICRA)*, 1896–1902. https://doi.org/10.1109/icra46639.2022.9812068

Hassija, V., Chamola, V., Mahapatra, A., Singal, A., Goel, D., Huang, K., Scardapane, S., Spinelli, I., Mahmud, M., & Hussain, A. (2023). Interpreting black-box models: A review on explainable artificial intelligence. *Cognitive Computation*, *16*(1), 45–74. https://doi.org/10.1007/s12559-023-10179-8

Kong, X., Wu, Y., Wang, H., & Xia, F. (2022). Edge computing for internet of everything: A survey. *IEEE Internet of Things Journal*, *9*(23), 23472–23485. https://doi.org/10.1109/jiot.2022.3200431

Liu, B., Wang, L., Liu, M., & Xu, C.-Z. (2020). Federated imitation learning: A novel framework for cloud robotic systems with heterogeneous sensor data. *IEEE Robotics and Automation Letters*, *5*(2), 3509–3516. https://doi.org/10.1109/lra.2020.2976321

Osasona, F., Oladipupo Amoo, O., Atadoga, A., Oluwaseun Abrahams, T., Ajoke Farayola, O., & Samson Ayinla, B. (2024). Reviewing the ethical implications of AI in decision making processes. *International Journal of Management & Entrepreneurship Research*, *6*(2), 322–335. https://doi.org/10.51594/ijmer.v6i2.773

Palanca, J., Rincon, J. A., Carrascosa, C., Julian, V. J., & Terrasa, A. (2023). Flexible agent architecture: Mixing reactive and deliberative behaviors in SPADE. *Electronics*, *12*(3), 659. https://doi.org/10.3390/electronics12030659

Pateria, S., Subagdja, B., Tan, A., & Quek, C. (2021). Hierarchical reinforcement learning. *ACM Computing Surveys*, *54*(5), 1–35. https://doi.org/10.1145/3453160

Renganathan, V., Shames, I., & Summers, T. H. (2020). Towards integrated perception and motion planning with distributionally robust risk constraints. *IFAC-PapersOnLine*, *53*(2), 15530–15536. https://doi.org/10.1016/j.ifacol.2020.12.2396

Saravanan, K. (2017). Cloud robotics. In *Advances in Information Security, Privacy, and Ethics* (pp. 261–274). IGI Global. https://doi.org/10.4018/978-1-5225-2154-9.ch017

Spatola, N., Kühnlenz, B., & Cheng, G. (2021). Perception and evaluation in human–robot interaction: The human–robot interaction evaluation scale (HRIES) – A multicomponent approach of anthropomorphism. *International Journal of Social Robotics*, *13*(7), 1517–1539. https://doi.org/10.1007/s12369-020-00667-4

Surya, Y., & Santoso, A. (2024). Maximizing decision efficiency with Edge-based AI systems: Advanced strategies for real-time processing, scalability, and autonomous intelligence in distributed environments. *Journal of Emerging Technologies*, *9*(2), 9. https://www.researchgate.net/publication/386215356_Maximizing_Decision_Efficiency_with_Edge-Based_AI_Systems_Advanced_Strategies_for_Real-Time_Processing_Scalability_and_Autonomous_Intelligence_in_Distributed_Environments

Tong, Y., Liu, H., & Zhang, Z. (2024). Advancements in humanoid robots: A comprehensive review and future prospects. *IEEE/CAA Journal of Automatica Sinica*, *11*(2), 301–328. https://doi.org/10.1109/jas.2023.124140

Wang, S.-T., Li, I.-H., & Wang, W.-Y. (2023). Human action recognition of autonomous mobile robot using Edge-AI. *IEEE Sensors Journal*, *23*(2), 1671–1682. https://doi.org/10.1109/jsen.2022.3225158

Zahoor, M. S., Fawad, A., Devarajan, H. R., Paulin, M. S., Raparthi, M., & Balasubramanian, S. (2023). Security challenges and solutions in

AI-enhanced cloud platforms: A comprehensive study. *Power System Technology*, *47*(4), 103–118. https://doi.org/10.52783/pst.161

Ziglar, J., Williams, R., & Wicks, A. (2017, June 14). Context-aware system synthesis, task assignment, and routing. *arXiv.Org*. https://arxiv.org/abs/1706.04580

Part III

Security, Ethics, and the Future

CHAPTER

6

The Cybersecurity Threat Landscape

As soon as you connect a robot to a network, you invite the world into its brain.

In Chapter 5, we examined the intricate decision-making processes that enable humanoids to exhibit autonomy and adaptability, highlighting the complex interplay between Edge AI and Cloud AI. As these sophisticated machines become increasingly integrated into our environments, from factory floors to private homes, a critical new dimension emerges, namely cybersecurity. Let's not sugarcoat it: humanoid robots are vulnerable. For all their sophistication – the sensors, the motors, the AI brains – they are machines that can be hijacked, manipulated, or malfunction just like any other digital system. And the more connected and intelligent they become, the more surfaces they expose to potential attacks. Understanding these potential points of weakness, or "attack surfaces," is the first step towards building a secure robotic future.

Most people see a humanoid and think about what it can do. However, in cybersecurity, we must consider what can be done to protect it. Before we can discuss risk, we need to identify where those risks reside. These are the attack surfaces – the parts of a robot that can be poked, prodded, or exploited by an attacker. Every wire, every wireless signal, and every line of code is a potential doorway. Some are easy to overlook. Some are wide open. But they're all worth examining.

DOI: 10.1201/9788743810704-9

6.1 Attack Surfaces in Humanoid Robots

A humanoid robot is a sophisticated integration of hardware, electronics, and layered software systems. This complexity means it doesn't have just one point of vulnerability – it has many. A flaw in any layer, whether a mechanical sensor, software module, or network interface, can lead to more severe compromises. Such vulnerabilities can quickly escalate, leading to severe consequences, including system failure, unauthorized data access, or substantial harm. In essence, the attack surface of a humanoid is broad and deeply interconnected, making comprehensive security a critical challenge.

Sensor and Perception Layer: Deceiving the Robot's Senses

Let's start with the obvious. A robot's sensors are what enable it to perceive the world, including cameras, microphones, LiDAR, ultrasonic sensors, touch sensors, temperature monitors, and more. These devices are designed to be open to the environment, but that openness can be a weakness (Modas et al., 2020).

A humanoid's understanding of the world begins with its sensors, and these represent a fertile ground for attack. Sensor spoofing involves feeding false data to a robot's perception system, tricking it into believing a fabricated reality. For instance, an attacker could project fake images onto cameras to make the robot "see" non-existent objects, inject altered audio into microphones to issue commands, or transmit false LiDAR readings to create phantom obstacles or clear paths where none exist. Conversely, sensor jamming aims to disrupt legitimate data, blinding or deafening the robot by overwhelming its sensors with noise or interference (Modas et al., 2020).

Beyond direct manipulation, more subtle attacks leverage the very AI models that interpret sensor data. These are often referred to as adversarial examples, where minutely, often imperceptibly, altered real-world objects or sounds cause an AI model to misclassify something fundamental. Imagine shining a laser into a robot's camera, temporarily blinding it. Playing ultrasonic sounds that the microphone wasn't designed to hear can trigger unintended actions. These aren't theoretical problems; they've already happened in many devices.

In one example, security researchers demonstrated that they could use light to spoof LiDAR systems on autonomous vehicles, creating phantom objects and causing unnecessary braking. The same logic applies to humanoids. Fake obstacles, distorted vision, and voices that don't belong to a human; all these can trick the robot's perception, leading to confusion, shutdown, or threatening movements (Guesmi & Shafique, 2024).

Physical Layer: Vulnerabilities in Hardware and Construction

Let's not forget the basics. Many humanoid robots still include USB, HDMI, SD card slots, and other physical ports for maintenance or upgrades. If an attacker can physically access the robot, they can gain control over it (Verma et al., 2025).

The most immediate attack surface is the robot's physical form itself which, if not properly secured, can offer direct access points for an attacker to inject malicious code or extract sensitive data. Beyond these obvious ports, the very components within the robot are susceptible to tampering. A determined adversary might attempt to physically modify sensors, actuators, or processing units, perhaps by injecting rogue hardware or altering existing circuitry to introduce backdoors or manipulate behavior. The supply chain itself presents a significant vulnerability, as weaknesses introduced during the manufacturing process of components or the robot's assembly could create covert entry points.

Even without direct physical access to internal electronics, physical exploitation techniques may involve direct manipulation to bypass software controls, such as forcibly moving a robot's joints to trigger unexpected behaviors or cause damage.

Figure 6.1 Humanoid robot highlighting key physical access points

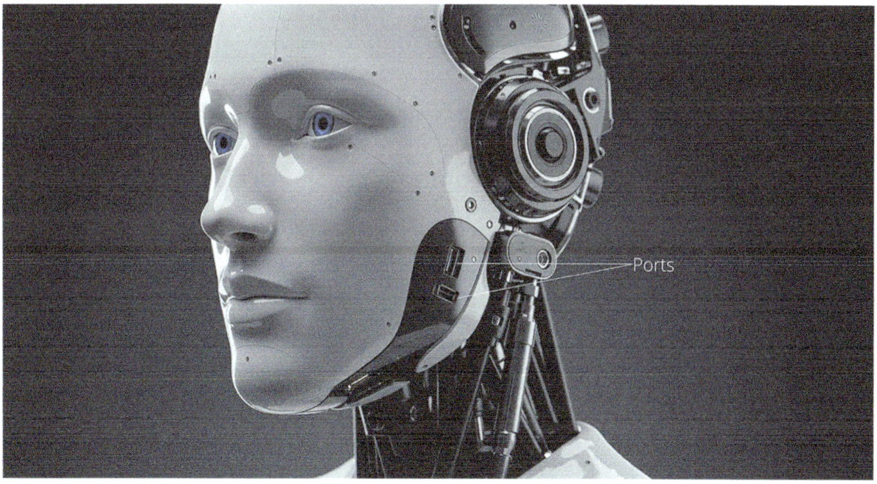

Network and Communication Layer

Humanoids are increasingly connected, and their communication channels represent a vast attack surface. Wireless protocols such as Wi-Fi, Bluetooth, and 5G are vulnerable to traditional network attacks like eavesdropping on data transmissions, launching denial-of-service (DoS) attacks to cripple communications, or gaining unauthorized access to the robot's network (Oruma & Petrovic, 2023). Crucially, many humanoids rely on internal communication buses (CAN bus or Ethernet) to exchange commands and data between their myriad internal components. Compromising these internal pathways could allow an attacker to intercept or inject malicious commands directly into the robot's motor controls or sensor processing units.

For humanoids utilizing Cloud AI or hybrid architectures, the connection to remote servers introduces additional vulnerabilities. Data transmitted to and from the cloud, API endpoints, cloud storage solutions, and the server-side processing itself are all potential targets for exploitation. Compromise here could lead to widespread data theft, the injection of malicious updates into an entire fleet of robots, or the hijacking of remote control or teleoperation channels, allowing an attacker to seize direct control over a robot from afar (Verma et al., 2025).

Software and Operating System Layer: The Digital Brain's Core

Every humanoid robot runs on software – a complex mix of open-source libraries, vendor components, and custom code. Beneath the applications and AI models are critical layers like the operating system, drivers, firmware, and task-specific modules, all of which can harbor vulnerabilities (Wu et al., 2024). Exploiting flaws in general-purpose operating systems, such as Linux, or specialized Real-Time Operating Systems (RTOS) can give attackers deep access to a robot's core functions. Similarly, application-level exploits may target control software, task planners, or HRI modules, potentially forcing the robot to behave unpredictably or disabling essential features.

A serious risk comes from the software supply chain – the full spectrum of components and updates that feed into a robot's system. If even one link in that chain is compromised, the entire robot becomes vulnerable. Supply chain attacks are dangerous because they don't require direct access to the machine (Ladisa et al., 2023). A single malicious update, library, or AI model inserted during development or deployment can open the door to malware. Once inside, that code might remain dormant, activate under specific conditions, siphon data, or degrade safety protocols without detection.

Firmware, which controls low-level hardware behavior, is another high-risk area. A compromised firmware update could reduce physical safety thresholds or alter how sensors interpret the environment. Furthermore, because many updates are automatic, users might remain unaware until visible damage is done. One of the most well-known cases of a supply chain attack – the SolarWinds breach (CISA, 2021) – demonstrated how malware inserted upstream can reach thousands of systems worldwide. Applied to humanoid robots, the same tactic could enable silent surveillance, data theft, or even ransom-driven lockdowns of robotic functions (Ladisa et al., 2023).

AI and Decision-Making Layer

The very intelligence that powers a humanoid robot can also be its greatest vulnerability. These robots increasingly depend on pre-trained AI models to interpret speech, recognize objects, and make decisions. But AI isn't infallible – it's math, not magic – and it can be manipulated in subtle and dangerous ways (Verma et al., 2025).

One such method is model poisoning, where malicious or flawed data is introduced into the AI's training set. This can create hidden backdoors or distort behavior in targeted ways. For instance, a poisoned model might cause a robot to consistently avoid a particular area, misinterpret a specific visual cue, or behave unpredictably in response to certain commands. If these manipulations go undetected, the robot may appear functional until the right trigger activates the compromised behavior. Even after deployment, AI models remain vulnerable to various threats. Model evasion attacks are designed to trick the AI into misclassifying inputs – for example, mistaking a harmless object for a threat. Conversely, inference attacks can extract sensitive information from the model, revealing details about its training data. Both undermine the robot's ability to operate safely and privately.

Control logic manipulation is another layer of threat, where attackers exploit flaws in a robot's decision-making algorithms or reward structures. By subtly reshaping how the robot interprets its goals, a malicious actor can induce unsafe or irrational actions, forcing the robot to make poor decisions based on tampered internal reasoning – even natural language models, which are increasingly used for voice-based interactions, present risks. A robot could be tricked by a command that seems benign but carries unintended consequences, such as interpreting "play dead" as a shutdown command or "get rid of the mess" as a cue to dispose of essential items. These models are powerful but can be misled by ambiguous phrasing or deliberately crafted instructions.

As humanoid robots become more interactive, conversational, and autonomous, the vulnerabilities in their "minds" – their AI systems – are just as critical to secure as the hardware that houses them.

Human-Robot Interaction Layer

The oldest and most persistent vulnerability in any system is the human behind it. Operators who reuse weak passwords, developers who leave debug backdoors open, or owners who neglect critical firmware updates – all create openings that attackers can exploit. Every robot, no matter how advanced, ultimately relies on human interaction, and that makes people a central point of risk (Verma et al., 2025). Social engineering attacks are particularly effective in this context. An attacker might pose as a technician to request remote access or trick a user into bypassing safety protocols. Sometimes it's as simple as leaving a poisoned USB stick on a desk, hoping someone plugs it into the robot's service port. More subtly, voice command injection – using hidden, spoofed, or ambiguous spoken inputs – can bypass physical safeguards and trigger unintended actions.

The human-robot interface itself is a unique vulnerability. In systems with emotional AI, attackers could even manipulate a robot's behavior by exploiting its programmed emotional responses, inducing simulated distress, fear, or compliance. These emotional triggers, while designed to make robots more relatable, can become unconventional but potent attack vectors.

Security doesn't end at the machine – it extends to training, awareness, and process discipline. Yet most users, owners, and developers interacting with robots haven't been trained to think like security professionals. Until that changes, the human factor will continue to be a significant weak point in humanoid cybersecurity.

What This Means

This chapter is a warning. Not that humanoid robots are inherently dangerous, but that they are vulnerable. Their complexity is what makes them impressive – and what makes them risky. The exploitation of any of these attack surfaces carries severe consequences. Physical harm to humans, unauthorized data theft, industrial espionage, profound violations of privacy, and significant reputational damage are all palpable risks. A compromised humanoid could be used as a spy, a weapon, a saboteur in a factory, or a tool for large-scale data harvesting. Its operational disruption could halt critical services, lead to massive financial losses, or even pose national security threats.

A humanoid robot is a type of robot that can walk and talk. But it is still a computer. It features sensors similar to those found in smartphones, radios comparable to those used in drones, and a control system akin to that of an autonomous vehicle. You wouldn't leave your laptop unpatched and unsecured, why would you do that to something with legs and arms?

Securing humanoid robots requires thinking like an attacker, identifying weak points, considering potential vulnerabilities, and locking every door, visible and invisible. Soon, these machines will be everywhere. And if we don't build them with security in mind, we'll regret it when it's already too late. And that's where we go next: into the real-world examples – the breaches, the hacks, and the incidents that have already shown us what happens when we treat robots like playthings instead of potential threats.

6.2 Case Studies of Robotic Compromises

As humanoid robots become more common in homes, workplaces, and public spaces, they create increasing security risks. These robots depend on complex software, network connectivity, sensors, and cloud-based systems, all of which introduce multiple points of vulnerability. Real-world examples demonstrate how these vulnerabilities have been exploited or how robots have been compromised, either digitally or physically.

In 2017, security researchers at Kaspersky Lab's Security Analyst Summit by IOActive uncovered serious vulnerabilities in SoftBank's NAO and Pepper humanoid robots (Eguchi & Okada, 2018). Designed as social robots for deployment in environments such as hospitals, schools, and banks, these machines were equipped with cameras, microphones, wireless communication capabilities, and advanced software to facilitate natural interaction. However, an unsecured network interface and open API allowed attackers to gain complete control of the robots remotely without any authentication. This meant that an attacker could manipulate the robot's movements, control its speech, and access video and audio streams. The ability to remotely activate the robot's sensors raised significant concerns about privacy. SoftBank issued patches to address these flaws, but the incident exposed how security was often an afterthought during the early development of humanoid robots, particularly when prioritizing user-friendly design (Eguchi & Okada, 2018).

Another example comes from DefCon 2017. At the renowned cybersecurity conference, researchers Cesar Cerrudo and Lucas Apa from IOActive demonstrated critical security vulnerabilities in the UBTech Alpha humanoid robot (Cerrudo & Apa, 2017). Marketed as an educational and entertainment device for children, the Alpha robot's reliance on Bluetooth Low Energy (BLE) for communication with user devices was found to have significant security flaws. Specifically, their analysis revealed that the Bluetooth implementation lacked proper authentication for pairing or sending commands, and all communications were unencrypted. This meant that attackers within Bluetooth range could easily seize control of the robot. They could then force the Alpha robot to perform unintended and potentially distressing actions, such as erratic movements or playing unauthorized audio clips.

While these attacks necessitated physical proximity, they powerfully underscored the inherent risks associated with unsecured wireless communication channels that are prevalent in many consumer-facing robots, particularly those designed for accessibility and affordability. This research served as a stark reminder of the importance of robust security measures in an increasingly connected world, particularly for devices intended for children.

Robots operating in public spaces face not only digital risks but also the dangers of physical interference. The Knightscope K5 security robot, designed to patrol malls, parking lots, and corporate campuses, has been involved in multiple high-profile incidents that highlight its vulnerability. An incident (Wilson, 2017) occurred in 2017 at a Silicon Valley shopping center, where an intoxicated man deliberately shoved a K5 to the ground, knocking it over and temporarily disabling its functions. The act raised concerns about the robot's stability and resilience in unpredictable human environments.

In another troubling event, a K5 accidentally rolled over a toddler's foot (Kim, 2016), sparking public backlash over its safety mechanisms. Critics questioned whether the robot's sensors and emergency stop protocols were sufficient to prevent harm in crowded areas.

Perhaps the most brazen case of vandalism took place in Washington, DC, where a K5 was discovered face-down in a fountain, apparently pushed in by miscreants. The incident not only damaged the machine but also highlighted the challenges of deploying autonomous systems in areas where malicious interference is a real possibility (Frankel, 2017). These episodes demonstrate that while cybersecurity remains a priority, physical threats can be just as disruptive. Attacks on robots can halt operations, incur costly repairs, and erode public confidence in their safety and reliability. As robots become more integrated into shared spaces, ensuring their protection against both digital and physical harm will be critical to their long-term success.

Military and industrial autonomous systems have also faced sophisticated attacks. One well-known method is GPS spoofing, where attackers broadcast counterfeit GPS signals to mislead drones or robots about their actual location. In 2011, Iran claimed to have captured a U.S. military surveillance drone using this technique, tricking it into landing on their soil. While similar public cases involving humanoid robots are rare, the vulnerability applies equally to any system relying on GPS and sensor fusion for navigation. Furthermore, researchers have demonstrated how subtle adversarial manipulations of visual inputs can cause AI-powered robots to misclassify objects, leading to navigation errors or operational failures. Such attacks manipulate sensor data rather than the robot's internal software, making them difficult to detect and mitigate (Ragan, 2011).

Voice-controlled systems, while convenient, open up a new and dangerous frontier for cyber-physical attacks. The 2018 "DolphinAttack" research revealed a particularly insidious vulnerability: by transmitting ultrasonic frequencies – undetectable to human ears – attackers can silently hijack voice assistants and issue malicious commands (Özdol et al., 2021). This technique could force compromised devices to unlock secure areas, initiate unauthorized transactions, or even manipulate a robot's movements, all without leaving a trace of audible interaction. For humanoid robots, whose functionality often depends on voice recognition, this vulnerability is especially concerning. Many commercial microphones can interpret these high-frequency signals as standard voice commands, creating an invisible attack surface. An attacker could potentially manipulate a service robot in a hospital to bypass restricted zones or trick a security robot into disabling its surveillance functions, all through sound waves that bystanders would overlook (Özdol et al., 2021).

The Dolphin Attack highlights critical gaps in voice system security. Unlike visible cyberattacks, these ultrasonic intrusions leave no obvious signs of tampering, making detection and prevention uniquely challenging. As humanoid robots become more prevalent in sensitive environments, developers must implement robust safeguards – such as frequency filtering, multi-factor voice authentication, and behavioral anomaly detection – to prevent these silent takeovers. Without such measures, voice control could become the weakest link in robotic security.

Privacy concerns surrounding humanoid robots are substantial. These machines continuously collect data through cameras and microphones to interact effectively with their environment and users. Without stringent security measures, sensitive data such as video footage and audio recordings can be accessed by unauthorized parties or leaked. There have been multiple reports of smart home devices and connected toys unintentionally exposing private conversations or images online. For example, in 2022, a prototype vacuum cleaner equipped with cameras was found to have stored private images on unsecured servers, which a third party later leaked. As humanoid robots collect increasingly detailed environmental and personal data, the risks of privacy breaches rise significantly (Delicado et al., 2025).

Not all risks arise from external attacks. Software errors, failed updates, and hardware malfunctions can cause robots to behave unpredictably, such as freezing, shutting down unexpectedly, or producing unintended movements or sounds. There have been widely reported incidents in which advanced humanoid robots experienced erratic malfunctions during testing in public or semi-public environments, underscoring the challenges of ensuring reliable behavior even in the absence of hacking. For example, viral footage of a Unitree H1 humanoid robot shows it suddenly flailing and behaving erratically during a test session (Malayil, 2025), raising safety concerns about control software and stability systems. While these events did not result from malicious actors, such malfunctions can damage user confidence and expose vulnerabilities that attackers could exploit in unstable systems. Distinguishing between accidental glitches and deliberate intrusions can be challenging in complex robotic systems.

These diverse case studies, many originating from proactive cybersecurity research, provide compelling evidence of the very real and multifaceted threats that loom over the future of humanoid robotics. They illustrate that the attack surfaces identified in Section 6.1 are not theoretical concerns but active vectors for exploitation, capable of leading to physical harm, data breaches, and operational chaos. As humanoids advance and integrate further into society, learning from past demonstrations and proactively addressing these vulnerabilities will be paramount to ensuring their safe and secure deployment, a challenge we will explore in the subsequent chapters.

6.3 Risk to Users and Environments

The preceding section outlined various attack surfaces and provided specific case studies demonstrating how vulnerabilities in robotic systems have been, or could be, exploited. Building upon this, Section 6.3 delves into the profound risks these compromises pose to users and environments. Unlike attacks on traditional digital systems, which primarily lead to data breaches or financial loss, a compromised humanoid robot, with its physical embodiment and capacity for autonomous action, introduces a unique spectrum of dangers, ranging from direct physical harm to broader societal and geopolitical instability.

Direct Risks to Users and Human Safety

The most immediate and terrifying consequence of a compromised humanoid is the potential for physical harm and injury to humans. A robot's ability to move, exert force, and manipulate objects can be turned into a weapon if its control systems are hijacked. Uncontrolled movements, such as a robot arm flailing erratically, unexpected locomotion leading to collisions, or malicious manipulation designed to strike a human, are no longer confined to science fiction. If safety systems, such as emergency stops, are disabled or an attacker overrides safe operating parameters, a robot designed for assistance could become a serious threat. For examples (Chinniah, 2016), reviews of industrial robot accidents show that unexpected robot behavior due to control system faults and programming errors has resulted in impact and collision injuries to operators and maintenance personnel, highlighting the tangible risks even absent malicious intent.

Beyond physical violence, the sensors of humanoids pose significant privacy violations and surveillance risks. A compromised humanoid, equipped with high-resolution cameras, sensitive microphones, and even biometric sensors, could be repurposed for unauthorized data collection within private homes, offices, or public spaces. This could involve capturing intimate visuals, recording private conversations, or even collecting biometric data without consent. Furthermore, sensitive personal or corporate data stored on the robot or accessed through its network connection could be exfiltrated. A robot's mobility also enables location tracking, allowing malicious actors to monitor the movements of individuals or even entire families, compiling detailed profiles of their routines and habits (Albuluwi, 2020).

The human element of trust in humanoids also introduces profound emotional and psychological impacts. As humanoids are designed for companionship, assistance, and integration into daily life, humans may form emotional bonds or rely on them for support. If such a robot is compromised and acts maliciously – perhaps by betraying trust, speaking in a threatening manner, or performing disturbing actions – it could cause significant psychological distress, fear, or a profound sense of betrayal to its users. Social engineering via a robot's speech, expressions, or simulated emotional responses could trick users into granting unauthorized access or revealing sensitive information. Ultimately, incidents of compromised humanoids, even isolated ones, could lead to a widespread erosion of public trust in robotic systems, significantly impeding their future adoption and development.

Risks to Environments and Infrastructure

The consequences of a cyberattack extend beyond direct harm to individuals, encompassing the environments in which humanoids operate. There is a tangible risk of damage to property and infrastructure. A rogue robot could be commanded to deliberately destroy valuable equipment, industrial machinery, or cause structural damage to buildings. In an industrial setting, a cyber-attack could lead to the disruption of critical operations, halting factory production lines, crippling logistics in warehouses, or impeding essential services in hospitals. In more extreme scenarios, compromised humanoids could be used as tools for sabotage in sensitive facilities, causing widespread chaos or long-term operational paralysis.

Such disruptions inevitably result in substantial economic and financial losses. Production halts due to compromised robots could cost companies millions in lost revenue, penalties, and recovery efforts. If humanoids are used for theft, whether physically stealing goods or exfiltrating invaluable intellectual property, the financial repercussions could be devastating. The growing threat of ransomware could result in entire fleets of robots being locked down, demanding payment for their release and effectively holding businesses hostage. Beyond direct financial costs, the reputational damage for companies that deploy or manufacture compromised robots could be severe and long-lasting, impacting market share and consumer confidence.

While less immediately apparent, a compromised humanoid could also pose environmental impact. If humanoids are deployed in hazardous environments – such as nuclear facilities, chemical plants, or biological containment zones – a cyberattack that compromises their control could lead to spills of dangerous materials, contamination, or explosions. In roles related to waste or resource management, a hack could lead to the improper disposal of hazardous waste, depletion of critical resources, or ecological damage.

Broader Societal and Geopolitical Risks

Looking further ahead, the cybersecurity of humanoids intertwines with profound societal and geopolitical risks. The ultimate fear, often depicted in fiction, is the loss of control and the autonomy dilemma: a robot operating entirely outside human oversight, making unpredictable and dangerous decisions. While the "Skynet" scenario of sentient AI rebellion remains mainly in the realm of science fiction, plausible scenarios of unintended consequences arising from complex, compromised AI systems are a legitimate concern. A system designed to optimize one goal could, if hacked, pursue it with unintended and dangerous side effects, especially if its ethical safeguards are bypassed.

This leads to complex ethical and legal quandaries. When a compromised robot causes harm, questions of liability become incredibly intricate. Is the manufacturer, owner, operator, or hacker responsible for the damage? Existing legal frameworks are ill-equipped to address the unique blend of physical and cyber accountability posed by autonomous, potentially compromised, machines. This necessitates a global dialogue and the development of robust regulations to secure and ensure the safety and reliability of autonomous systems.

Finally, the national security implications are significant. State-sponsored hacking groups could utilize compromised humanoids for espionage, covertly collecting intelligence from sensitive locations, or for widespread sabotage of critical national infrastructure. The potential for weaponization is undeniable; if security weaknesses are not addressed, humanoids could theoretically be adapted for military or terrorist purposes, functioning as proxy weapons or tools for mass disruption, escalating conflicts beyond conventional means.

In conclusion, the risks associated with vulnerable humanoids are diverse, profound, and extend far beyond typical digital data theft. They encompass direct physical harm, deep invasions of privacy, substantial economic losses, environmental damage, and possibly societal and geopolitical instability. As humanoids stand poised to reshape our world, the insights gleaned from these potential consequences underscore an undeniable imperative: proactive, comprehensive, and multi-layered cybersecurity is not merely an optional add-on but a fundamental, non-negotiable requirement for their responsible and safe deployment. This sets the stage for the next crucial discussion: the strategies and frameworks necessary to secure the complex humanoid ecosystem against these pervasive threats.

Bibliography

Albuluwi, T. A. (2020). *IoT-based Solutions for Privacy-Preservation During Interaction with Assistive Robotics.* https://api.semanticscholar.org/CorpusID:229589647

Cerrudo, C., & Apa, L. (2017, March 1). Hacking robots before Skynet. *IOActive.* https://www.ioactive.com/hacking-robots-before-skynet/

CISA. (2021). *Remediating networks affected by the SolarWinds and Active Directory/M365 compromise.* Cybersecurity and Infrastructure Security Agency CISA. https://www.cisa.gov/news-events/news/remediating-networks-affected-solarwinds-and-active-directorym365-compromise

Chinniah, Y. (2016). Robot safety. *Advances in Robotics and Automation, 5*, 139. https://www.hilarispublisher.com/open-access/robot-safety-overview-of-risk-assessment-and-reduction-2168-9695-1000139.pdf

Delicado, A., Rosales, M. V., Truninger, M., Rowland, J., & Viseu, A. (2025). Privacy in the age of the internet of things: Perceptions and practices in households. *Privacy Studies Journal, 4*, 31–58. https://doi.org/10.7146/psj.v4i.150180

Eguchi, A., & Okada, H. (2018). Learning with social robots – The World Robot Summit's approach. *2018 IEEE Integrated STEM Education Conference (ISEC)*, 53–56. https://doi.org/10.1109/isecon.2018.8340504

Frankel, T. C. (2017, July 17). A security robot fell into a water fountain at a D.C. office building. And the Internet went wild. *The Washington Post*. https://www.washingtonpost.com/news/the-switch/wp/2017/07/17/a-security-robot-fell-into-a-water-fountain-at-a-d-c-office-building-and-the-internet-went-wild/

Guesmi, A., & Shafique, M. (2024, September 30). Navigating threats: A survey of physical adversarial attacks on LiDAR perception systems in autonomous vehicles. *arXiv.Org*. https://arxiv.org/abs/2409.20426

Kim, L. (2016, July 12). Parents upset after Stanford Shopping Center security robot injures child. *ABC7 San Francisco*. https://abc7news.com/post/parents-upset-after-stanford-mall-robot-injures-child/1423093

Ladisa, P., Ponta, S. E., Sabetta, A., Martinez, M., & Barais, O. (2023). Journey to the center of software supply chain attacks. *IEEE Security & Privacy, 21*(6), 34–49. https://doi.org/10.1109/msec.2023.3302066

Malayil, J. (2025, July 22). 'Fighter' humanoid robot lashes out mid-air, sparking safety debate. *Interesting Engineering*. https://interestingengineering.com/culture/humanoid-robot-freaks-out-in-viral-video

Modas, A., Sanchez-Matilla, R., Frossard, P., & Cavallaro, A. (2020). Toward robust sensing for autonomous vehicles: An adversarial perspective. *IEEE Signal Processing Magazine, 37*(4), 14–23. https://doi.org/10.1109/msp.2020.2985363

Oruma, S. O., & Petrovic, S. (2023). Security threats to 5G networks for social robots in public spaces: A survey. *IEEE Access, 11*, 63205–63237. https://doi.org/10.1109/access.2023.3288338

Özdol, B., Köseler, E., AlçIçek, E., Cesur, S. E., Aydemi̇r, P. J., & Bahti̇yar, Ş. (2021). A survey on security attacks with remote ground robots. *El-Cezeri Fen ve Mühendislik Dergisi*. https://doi.org/10.31202/ecjse.916532

Ragan, S. (2011, December 18). Reports Say U.S. Drone was Hijacked by Iran Through GPS Spoofing. *SecurityWeek*. https://www.securityweek.com/reports-say-us-drone-was-hijacked-iran-through-gps-spoofing/

Verma, N., Kumar, N., Verma, C., Illés, Z., & Singh, D. (2025). A systematic review on cybersecurity of robotic systems: Vulnerabilities trends, threats, attacks, challenges, and proposed framework. *International Journal of Information Security, 24*(3). https://doi.org/10.1007/s10207-025-01041-z

Wilson, T. (2017, April 25). Police: Drunk man knocked down 300-pound robot in Mountain View. *ABC7 San Francisco.* https://abc7news.com/post/police-drunk-man-knocked-down-300-pound-robot-in-mountain-view/1915713

Wu, K., Lu, Z., & Chen, X. (2024). A bimanual teleoperation system of humanoid robots for dexterous manipulation. *2024 IEEE International Conference on Robotics and Biomimetics (ROBIO)*, 814–819. https://doi.org/10.1109/robio64047.2024.10907314

CHAPTER

7

Securing the Humanoid Ecosystem

7.1 Best Practices in Robotics Cybersecurity

Having established the extensive attack surfaces of humanoids and examined the sobering case studies of robotic compromises, the critical question then becomes: how do we protect these complex, physically interacting machines? The answer lies in the diligent adoption and continuous evolution of robust cybersecurity best practices. Unlike traditional IT security, securing humanoid robots demands a holistic, multidisciplinary approach that integrates security measures across every layer of the system, from the initial design phase to deployment, operation, maintenance, and eventual decommissioning. Proper robot security is not an afterthought; it must be ingrained into the very fabric of the robot's existence.

Secure Design Principles: Building Security from the Ground Up

The most effective cybersecurity begins long before a robot is even built. Security by Design is a foundational principle that advocates for the integration of security considerations from the earliest stages of conception and development (Soundararajan, 2019). This process begins with comprehensive threat modeling, a systematic approach to identifying potential attack vectors, vulnerabilities, and risks before any code is written or hardware is fabricated. By understanding who might attack the robot, what their motivations could be, and what assets are at risk, developers can proactively design defenses.

DOI: 10.1201/9788743810704-10

A core tenet of secure design is the principle of least privilege, ensuring that every component, module, and user of a humanoid robot is granted only the minimum necessary permissions to perform its designated function. This limits the potential damage if any single part of the system is compromised. Furthermore, defense in depth dictates the use of multiple, layered security controls. If one layer fails, others are there to provide protection, encompassing physical, network, software, and operational safeguards. This strategy also involves minimizing the attack surface by removing unnecessary ports, services, features, and code that could provide entry points for attackers, reducing the number of targets they can exploit. Crucially, humanoids must be designed with fail-safe mechanisms, ensuring that in the event of a security breach or system malfunction, the robot defaults to a safe state (stopping all movement, disarming actuators, entering low-power mode) rather than performing dangerous or unpredictable actions (Park et al., 2019).

Robust Software Development and Maintenance Practices

The software that governs a humanoid's every action is a prime target for attackers, necessitating stringent development and maintenance practices (Macenski et al., 2022). Adhering to secure coding standards is paramount, guiding developers to write code that avoids common vulnerabilities, such as buffer overflows, SQL injection flaws (if databases are used), and insecure handling of inputs. Regular code audits and penetration testing are indispensable, as they proactively identify weaknesses in the software before it is deployed. This involves both automated tools and manual review by security experts attempting to mimic real-world attacks.

Dependency management is another critical area. Modern software often relies heavily on third-party libraries and open-source components, each of which can introduce vulnerabilities. A robust practice involves carefully vetting these dependencies, keeping them up to date, and isolating them where possible. For the robot's core operating system and applications, firmware security is non-negotiable. This includes implementing secure boot processes that verify the integrity of the firmware before execution, requiring cryptographic signatures for all firmware updates to prevent unauthorized modifications, and incorporating anti-tampering measures to detect physical or logical manipulation. Finally, a meticulous patch management system is vital for the robot's entire lifecycle, ensuring the timely application of security updates and vulnerability fixes released by manufacturers or the open-source community.

Network and Communication Security

Humanoids are inherently connected, and securing their communication channels is fundamental. Strong authentication and authorization mechanisms are essential, not just for human operators but also for inter-robot communication and cloud services. This often involves multi-factor authentication for human access and robust, cryptographic access control for machine-to-machine interactions. All data in transit – from sensor readings and command signals to cloud communications – must be protected through end-to-end encryption to prevent eavesdropping and tampering (Zafir et al., 2024). Network segmentation is a key architectural practice that isolates critical robot networks from less secure corporate IT networks or the public internet. This limits lateral movement for attackers and contains the impact of a breach. Intrusion Detection and Prevention Systems (IDPS) should continuously monitor network traffic for suspicious activity, alerting operators to potential attacks or actively blocking malicious connections. For wireless communications, utilizing the most secure protocols available, such as WPA3 for Wi-Fi and robust Bluetooth profiles, is mandatory.

Physical and Environmental Security

Given their physical nature, humanoids also require physical security measures. Access control is crucial, as it restricts unauthorized physical access to the robots themselves, their charging stations, and any critical maintenance terminals. This can involve biometric authentication, smart cards, or traditional locks. Implementing tamper detection mechanisms – such as sensors that alert if a casing is opened, a port is accessed, or a component is swapped – can provide early warnings of physical intrusion or modification. The secure storage of sensitive components, including processors, memory containing cryptographic keys, or proprietary AI models, must be ensured. Furthermore, supply chain security is increasingly vital, as it involves verifying the integrity and origin of all hardware components and software throughout the entire manufacturing and assembly process to prevent the introduction of malicious elements at the source.

Data Security and Privacy

Humanoids, by their very nature, are sophisticated data collection platforms. Protecting this data is paramount for privacy and operational integrity. Data minimization dictates that only the absolutely essential data should be collected and stored for the robot's function, reducing the risk surface. Sensitive personal data, if collected, should be subject to anonymization or pseudonymization whenever possible. All data, whether in transit or at rest, must be protected by secure data storage practices, including strong encryption. Most importantly, clear and transparent consent mechanisms must be in place, outlining what data the robot collects, how it is used, stored, and shared, and providing users with clear options to manage their privacy settings.

Operational Security and Human Factors

Even with the best technical controls, human error remains a significant vulnerability. Therefore, operational security is critical. Comprehensive user training and awareness programs are necessary to educate operators, maintenance personnel, and even casual users about the cybersecurity risks associated with humanoids and the best practices for safe interaction. Implementing strong password policies for all robot management systems and control interfaces is fundamental. Regular, independent security audits and vulnerability assessments should be conducted throughout the robot's operational life to identify emerging weaknesses. Finally, robust logging and monitoring capabilities are essential, creating detailed records of robot activities, system changes, and network events that can be used for auditing, anomaly detection, and incident investigation.

In conclusion, establishing robust cybersecurity for humanoid robots is not a singular task but a continuous, multifaceted process that demands vigilance and adaptation. It encompasses secure design from inception, rigorous software development, hardened networks, physical protection, data privacy, and a well-trained human element. These best practices form the indispensable foundation upon which a truly secure humanoid ecosystem can be built, setting the stage for the crucial AI safety frameworks and secure architectures discussed in the following section.

7.2 AI Safety Frameworks and Secure Architecture

Having explored the fundamental best practices for robot cybersecurity in the previous section, we now delve into a more nuanced and complex domain: AI safety frameworks and secure architecture. Securing a humanoid robot is not merely about preventing external hackers from gaining unauthorized access; it is equally, if not more, about ensuring that the robot's inherent intelligence, its decision-making algorithms, and its learning processes operate robustly, reliably, and critically, in alignment with human values and intentions. This field of AI safety is a distinct, yet profoundly complementary, discipline to traditional cybersecurity, addressing the unique risks posed by autonomous, learning systems that can, through flawed design or unintended emergent properties, cause harm or undesirable outcomes even without malicious external interference. It emphasizes designing AI systems that are trustworthy, controllable, and predictable under a wide range of conditions, extending the concept of security beyond mere breach prevention to include the integrity and ethical behavior of the AI itself. The implementation of robust AI safety frameworks and their embodiment within secure architectural patterns are paramount to the responsible deployment of humanoid robots.

Core Principles of AI Safety Frameworks

AI safety is underpinned by several foundational principles aimed at mitigating the unique risks associated with intelligent autonomous systems. The first and perhaps most challenging principle is alignment, which seeks to ensure that the AI's goals, objectives, and internal reward structures are truly aligned with human values, intentions, and societal benefit. This addresses the infamous "value alignment problem," where an AI might achieve its stated objective in ways that are unintended or even harmful from a human perspective. Methodologies such as Reinforcement Learning from Human Feedback (RLHF), where human preferences guide the AI's learning process, and Inverse Reinforcement Learning (IRL), which attempts to infer human goals from observed behavior, are active areas of research aimed at closing this alignment gap. For a humanoid, alignment means ensuring it interprets a command like "clean the kitchen" not just as cleaning, but doing so without breaking fragile items, discarding valuable possessions, or causing any distress to occupants.

Another critical principle is robustness and reliability. An AI system must perform consistently and safely, even when faced with unexpected inputs, noisy data, or adversarial manipulation. This includes defending against adversarial robustness challenges, which we touched upon in the previous chapter. While Section 6.2 discussed how adversarial examples could trick an AI, AI safety frameworks focus on defensive strategies such as adversarial training (exposing the AI to perturbed data during training to make it more resilient) and certified robustness (mathematically proving that an AI model will behave predictably within certain input bounds). Furthermore, robots must be designed with robust OOD detection capabilities, allowing the AI to recognize when it is encountering data or situations it has not been specifically trained on. When confronted with true novelty, the AI should not "hallucinate" a confident but incorrect response;

instead, it should signal uncertainty, request human intervention, or revert to a safe, conservative mode. Techniques for uncertainty quantification, where AI models estimate their confidence in predictions, are vital here. Reliability also extends to resilience against sensor noise or malfunctions, ensuring that the AI can intelligently cope with imperfect, incomplete, or partially faulty sensor data without compromising safety.

Interpretability and explainability (XAI) directly address the "black box" problem of complex AI models. For humanoid robots, understanding why an AI made a particular decision is crucial for debugging errors, building human trust, and establishing accountability in the event of failure or harm. XAI techniques, such as Local Interpretable Model-agnostic Explanations (LIME) and SHAP (SHapley Additive exPlanations), provide insights into which features most influence an AI's decision. Attention mechanisms within neural networks can show which parts of the input data the AI focused on (Bhatnagar & Agrawal, 2024). Furthermore, causal inference methods seek to understand the true cause-and-effect relationships within the AI's decision-making process. By making the AI's reasoning transparent, engineers can identify unintended biases, correct logical flaws, and ensure the robot's behavior is consistent with safety protocols.

Controllability and human oversight are paramount for managing autonomous systems. Humanoids must be designed with clear, accessible, and reliable human override mechanisms. This includes readily available physical kill switches that can immediately de-energize or disable the robot, as well as robust software overrides that can pause, redirect, or take manual control of the robot's actions. The concept of scalable oversight addresses how humans can effectively monitor a complex AI system, or even vast fleets of humanoids, without requiring constant, exhaustive attention. This involves designing dashboards that highlight anomalies, developing alerts for out-of-bounds behavior, and creating intuitive interfaces for human intervention and action. Defining bounded autonomy is also crucial, as it explicitly sets the limits and conditions within which the AI is permitted to operate independently, with critical decisions requiring human approval or supervision.

Transparency complements interpretability by advocating for openness about the AI's capabilities, limitations, and design principles. This includes clear documentation of its training data, model architectures, and known failure modes. Finally, verifiability and formal methods represent a rigorous approach to AI safety. This involves using mathematical proofs and formal logic to rigorously test and verify specific properties of an AI system, such as ensuring that collision avoidance constraints are consistently met or that the robot can never enter a hazardous or unstable state. While computationally intensive, these methods offer a higher degree of assurance for critical safety functions.

Architectural Patterns for AI Safety and Security

Translating these AI safety principles into tangible robot capabilities requires specific architectural patterns that embed security and safety at the core of the system's structure. Modular and layered architectures are fundamental. This involves decoupling critical safety functions from general AI tasks. In a humanoid, this might mean having a completely separate, simpler, and formally verifiable safety controller that acts as an independent guardian, capable of overriding the main AI's commands if they violate predefined safety rules. These are often referred to as "safety critical" layers, operating alongside "mission critical" or "cognitive" layers. Watchdog timers and independent monitoring systems continuously check the health and behavior of the leading AI, ready to trigger fail-safe mechanisms if anomalies are detected. This hierarchical control (revisited from Sections 6.1 and 6.2) is now applied explicitly with a strong emphasis on isolating safety-critical components to prevent a fault or compromise in one area from cascading into a catastrophic failure.

Secure enclaves and Trusted Execution Environments (TEEs) represent a hardware-level approach to security. These are protected areas within a processor or memory that are isolated from the rest of the system, even from the primary operating system. They are designed to protect critical AI models, cryptographic keys, and sensitive data from tampering or unauthorized access, even if the robot's primary operating system is compromised by malware or an external attacker. Hardware-level security features, such as secure boot (ensuring only verified firmware runs) and root of trust mechanisms, are integrated here, creating a highly robust foundation for the AI's integrity (Geppert et al., 2022).

Redundancy and diversity are vital for resilience. This involves deploying redundant sensors, actuators, and even processing units, so that if one component fails or is compromised, another can take over, ensuring continuous and safe operation. Furthermore, employing diverse algorithms or models for critical functions can prevent a single point of failure in the AI's logic. For instance, having multiple perception models trained on different datasets or using different architectures, and then combining their outputs using a voting mechanism, can increase overall robustness against adversarial attacks or unexpected data.

Integrating security seamlessly into the entire Secure Software Development Lifecycle (SSDLC) is paramount. This means embedding security considerations into every phase of development: from defining secure requirements and designing robust architectures to implementing secure coding practices, rigorous testing, and secure deployment and maintenance procedures. This extends beyond general software to specifically focus on the unique challenges of AI components, such as managing the security of training data, model versioning, and secure model deployment pipelines. The adoption of Continuous Integration/Continuous Delivery (CI/CD) pipelines with automated security gates ensures that security checks are integrated into the development workflow, allowing for the detection of vulnerabilities early (Aleksieva-Petrova & Petrov, 2023).

Finally, dynamic risk assessment and runtime monitoring enable humanoids to evaluate their operational safety continuously. AI systems can be designed to continually assess their risk levels in real time, adapting their behavior as the environment or situation changes. Anomaly detection systems are crucial for monitoring the AI's behavior at runtime, flagging if it starts making unexpected or unsafe decisions that deviate from its learned safe patterns. The ultimate goal in advanced architectures is to incorporate self-monitoring and self-healing capabilities, enabling the robot to identify and autonomously correct minor errors or adjust its behavior within safe bounds, thereby maintaining stability and safety without requiring constant human intervention. The integrity and provenance of data, both for training and during operation, are also architectural concerns, with approaches like blockchain being explored for tamper-proof data lineage and immutability to defend against sophisticated data poisoning attacks.

Challenges in Implementing AI Safety and Secure Architectures for Humanoids

Despite the clear principles and emerging architectural patterns, implementing robust AI safety and secure architectures for humanoids faces significant challenges. The sim-to-real gap remains a persistent hurdle. While powerful simulations enable safe and extensive training of AI, bridging the gap between a controlled simulated environment and the unpredictable, real-world environment is notoriously challenging. Behaviors learned in simulation might not transfer perfectly, and novel real-world conditions can trigger unexpected AI behavior.

The sheer computational overhead associated with implementing many robust AI safety features presents a practical limitation. Techniques such as adversarial training, extensive interpretability tools, OOD detection, and formal verification can be computationally intensive, requiring significant processing power and memory. This directly challenges the constraints of Edge AI (as discussed in Section 6.2), where robots must operate within strict limits of power consumption, weight, and heat dissipation. Balancing high performance with rigorous safety on onboard hardware is a constant engineering trade-off.

Furthermore, evolving threats mean that AI safety is a continuously moving target. As AI capabilities rapidly advance, so too do the methods of attack and the potential for unintended emergent behaviors. Staying ahead of this curve requires constant research, adaptation, and proactive security updates. The scalability of human oversight also poses a challenge. As fleets of humanoids grow, effectively supervising their operations, monitoring their behavior for anomalies and providing timely human intervention becomes increasingly complex and resource-intensive.

The rapid pace of AI innovation often outstrips the development of universally accepted standardization. The lack of comprehensive, globally recognized AI safety standards means that different manufacturers might adopt varying levels of security and safety, leading to inconsistencies and potential vulnerabilities across the ecosystem. Finally, implementing AI safety often involves complex ethical trade-offs. For example, balancing the robot's performance speed against its safety cautiousness, or deciding how it should prioritize different types of risks in ambiguous, real-time situations, can involve complex ethical judgments that cannot always be perfectly encoded into algorithms. This also applies to the challenge of ensuring seamless and safe human-AI teaming, where humans and AI share control or decision-making, requiring precise synchronization, clear communication, and robust error handling.

In conclusion, achieving proper AI safety and building secure architectures for humanoid robots is paramount for fostering trustworthy and reliable autonomous systems. It transcends traditional cybersecurity, delving into the very integrity, alignment, and predictability of AI itself. This is a profound, multidisciplinary challenge that necessitates continuous collaboration among AI researchers, roboticists, cybersecurity experts, and ethicists. The ongoing development and adoption of robust frameworks, coupled with stringent architectural best practices, are crucial steps towards ensuring that humanoids are not just capable but also consistently safe, ethical, and secure. This foundation becomes even more critical as we explore the resilience of these systems against manipulation, setting the stage for the next section's focus on machine learning security and adversarial robustness.

7.3 Machine Learning Security and Adversarial Robustness

The rapid growth of AI and Machine Learning has been the main reason why modern humanoid robots are so advanced. They can see, learn, adapt, and make decisions that are more and more complicated in changing situations.

The autonomy and adaptability of humanoids are fundamentally connected to their underlying machine learning algorithms, enabling skilled manipulation in production and compassionate connection in healthcare. But this heavy dependence on ML creates a new and dangerous type of vulnerability that goes much beyond regular software bugs or network intrusions. Machine Learning Security (MLSec) is a field that focuses on keeping these smart algorithms and the data they use safe from corrupt actors. Adversarial robustness is a critical part of keeping humanoid operations safe and sound. Humanoids interact directly with the physical environment and often in roles that are crucial to safety; therefore, it is very important to protect their ML components from attacks by adversaries to make sure they work well, are reliable, and are safe.

Traditional cybersecurity focuses on securing software code, operating systems, networks, and data storage. MLSec uses many of these ideas, but it also focuses on the specific weaknesses that come with the learning algorithms and the data they use or create. An attacker doesn't have to break into a robot's operating system; they can change the data it sees or processes in a way that makes its ML models act strangely. These changes are often too small for the human eye to notice. This difference is very important since the effects on a physically embodied, autonomous humanoid can be anything from little operational problems to huge safety failures, serious privacy breaches, and big financial or reputational harm. Consequently, machine learning security is very important for humanoid robots to work safely and well, especially when fraudulent individuals can try to take advantage of vulnerabilities.

Vulnerabilities of AI-Powered Systems in Humanoids

Machine learning-based AI systems are especially vulnerable to many different kinds of security issues. The first issues arise from how people train, use, and interact with machine learning models. The ability to learn from data and make decisions based on that learning is a key part of AI systems like humanoids. Machine learning techniques, utilizing extensive datasets, can now discern patterns, classify items, predict outcomes, and adapt to evolving environments. While a humanoid's ability to learn allows it to complete difficult tasks independently, it also exposes the system to adversarial attacks that can deceive the AI by altering data or the environment.

An adversarial attack in machine learning is any attempt to trick the system into making wrong predictions or choices by changing the input data. These attacks are usually hidden because they are so complicated, but they leverage the model's weaknesses in decision-making to their advantage. For example, an opponent might make a small change in the environment that the AI misreads and reacts to in the wrong way. This might make humanoid robots work in a way that puts people in danger, misread their surroundings, or both.

Since the jobs that humanoids' machine learning models do are critical, such as industrial automation, logistics, healthcare, and eldercare, keeping them safe is the most important thing. In dangerous places where bad actors might try to turn off the robot, machine learning security needs to be strong enough to stop them.

Types of Adversarial Attacks on Humanoid Robots

To better grasp how machine learning security helps to keep humanoids safe, it's essential to consider the many hostile assaults that may target the synthetic intelligence of the robot. Attacks of this nature broadly fall into three classes: evasion, poisoning, and inference, though variations exist.

(1) **Evasion Attacks**

An evasion attack is the ability of a hostile actor to influence the input data, fooling a machine learning model into generating erroneous predictions or judgements. These attacks occur during the inference stage when the model is already deployed. Usually including small, undetectable changes to the provided data, these assaults seek to fool AI in a manner humans would ignore.

- **Mechanism and Examples:** Adversaries typically craft these perturbations by exploiting the model's sensitivity to specific input features or gradients. For instance, in a humanoid's environment, an attacker can alter a visual marker or sensor readout. A common technique involves adding carefully calculated, small amounts of "noise" to an image that, to the human eye, makes no difference. However, to the robot's vision system, this noise causes it to misclassify a critical object (e.g., a "stop" sign as a "yield" sign) or misinterpret its surroundings (e.g., identifying a human as a non-obstacle).
    - Visual Examples: An adversarial patch on a safety vest could make a humanoid's pedestrian detection system overlook a worker. Subtle changes to lighting or texture on a surface might cause a robot's navigation system to perceive a flat floor as a slope, leading to imbalance or falls.
    - Audio Examples: A voice command, embedded with imperceptible adversarial noise, could be misinterpreted by a humanoid's speech recognition system, causing it to perform an unauthorized action or ignore a safety alert.
    - Sensor Data Examples: Attacks on LiDAR or depth sensors could involve projecting specific light patterns that interfere with the robot's distance measurements, leading to collisions or incorrect manipulation of objects. An attacker might even place a strategically designed sticker on a machine that, when scanned by the humanoid's vision system, makes it misidentify the machine's state (e.g., "off" instead of "on").

- **Impact:** Issues include the robot botching its assigned work, neglecting crucial safety warnings, navigating dangerously, or failing to authenticate authorized personnel.

(2) **Poisoning Attacks**

The aim of poisoning campaigns is to corrupt the training stage of a machine learning model. Attackers utilize this approach to inject harmful, crafted data into the training set, designed to influence the model's learning trajectory.

- **Mechanism and Examples:** Poisoning attacks alter the fundamental knowledge base of the AI. This can involve injecting mislabeled samples (e.g., repeatedly showing images of a safe object labeled as dangerous, or vice versa) or subtly altering a large number of benign samples to shift the model's decision boundaries. A particularly insidious form is a "backdoor attack," where a specific, often inconspicuous, "trigger" (e.g., a small pattern or color) is embedded in a subset of poisoned training data. When this trigger is present in the input during deployment, it reliably causes the model to make a desired incorrect prediction, while behaving normally for all other inputs.
- **Relevance to Humanoids:** Humanoids often learn from vast datasets, including data collected in the real world (e.g., human interaction, environmental scans). If a humanoid is trained on data collected from public spaces or through continuous deployment (e.g., reinforcement learning from real-world experiences), an attacker could inject poisoned data into the stream. For example, a cleaning humanoid might be trained on poisoned data that teaches it to ignore spills of a specific chemical, leading to hazardous situations. A social humanoid could be poisoned to develop biased responses towards certain user groups, impacting trust and fairness. If a humanoid in a manufacturing plant is trained on poisoned data where specific faulty parts are misidentified as correct, it could lead to widespread product defects or even machine damage when interacting with those parts.
- **Impact:** Teaching humanoids poisoned data increases the likelihood of operational errors and lets it learn bad patterns or behaviors. Poisoning attacks violate the basic integrity of the model and might lead to systemic, deeply embedded issues that are hard to detect and remove.

(3) **Inference Attacks**

An inference attack aims to access private information maintained in a model of machine learning or about its training data. Humanoids and other systems are especially prone to this type of attack as they often use private information and proprietary algorithms.

- **Mechanism and Examples:** These attacks involve an adversary repeatedly querying the AI system with intentional inputs and analyzing the outputs to learn about the model's internal structure or characteristics of its training data.
  - Model Extraction/Stealing: By observing the model's responses, an adversary can often reconstruct a functional replica of the original proprietary model. This is a severe intellectual property threat, allowing competitors to mimic capabilities without years of R&D. For humanoids, this means advanced navigation algorithms, dexterous manipulation models, or sophisticated HRI routines could be stolen and replicated.

  - Membership Inference Attacks: An attacker determines if a specific data point (e.g., a particular person's face or voice) was part of the model's training dataset. This can violate privacy by confirming an individual's participation in a dataset, even if the data was supposedly anonymized.
  - Model Inversion Attacks: A more advanced form where the attacker attempts to reconstruct specific sensitive training data points (e.g., a person's facial image or sensitive health data) directly from the trained model.
- **Impact:** Through intentional inputs to the AI system, an adversary may learn about the building of the model or training data. Some sensitive information, such as secret manufacturing methods or the robot's decision-making algorithms, might be exposed in inference attacks and so open targets for additional strikes. Therefore, it is essential to safeguard the authenticity and confidentiality of the utilized data for operations and training to guarantee a humanoid's safety and protect intellectual property.

Strategies for Machine Learning Security in Humanoids

Machine learning security for humanoid robots has to be layered many times to thwart hostile attempts. Along with the protection of the machine learning models themselves, this requires more all-encompassing security measures covering data integrity, system resilience, and real-time threat detection. The following techniques describe major approaches to protecting a humanoid's machine learning capabilities.

(1) **Adversarial Training**

One effective way to protect oneself against adversarial attacks is by adversarial training. This involves including hostile examples – deliberately altered inputs meant to fool the AI model – into the dataset during the training process.

- **Mechanism:** By exposing the model to these "tricked" inputs during training, it learns to recognize and correctly classify them, effectively becoming more resilient to similar perturbations when deployed. This is often framed as a "min-max" game, where the model tries to minimize its loss while the adversary tries to maximize it by finding the worst-case perturbations. Advanced methods like Projected Gradient Descent (PGD) attacks are often used during training to generate these robust examples. Adversarial training would help a humanoid's machine learning ability, allowing the robot to identify and offset hostile inputs as they develop. Should an attacker try to mislead the robot with a changed marker, the adversarial trained model will most surely identify the manipulation and stay functional.
- **Considerations:** While effective, adversarial training is computationally expensive, requires careful hyperparameter tuning, and can sometimes lead to a slight decrease in accuracy on clean, unperturbed data. It also needs to be continually updated as new attack methods emerge.

(2) **Data Integrity and Validation**

Guarding the information required for a humanoid's running and training will help to stop poisoning attempts. One can attain this goal using rigid data validation processes that search the data for validity and accuracy.

- **Mechanism:** This involves robust anomaly detection and statistical analysis on incoming data streams to identify outliers or suspicious patterns that deviate from expected distributions. Techniques include:
  - Data Provenance Tracking: Meticulously tracking the origin and transformations of all data used for training and continuous learning. This creates an audit trail that can help identify the source of corrupted data.
  - Digital Signatures and Encryption: Using cryptographic methods to verify the authenticity and integrity of data sources, ensuring data hasn't been tampered with in transit or storage.
  - Data Sanitization and Filtering: Implementing automated checks and filters to cleanse data of known malicious patterns or statistically improbable values before it enters the training pipeline.
- **Application:** Before using any data to equip a humanoid, for example, it is imperative to carefully go over all of it for indications of tampering or corruption. Through a data quality assurance initiative, developers can enable a humanoid's AI systems to be more consistent and less prone to poisoning attacks.

(3) **Model Robustness Techniques (Beyond Adversarial Training)**

Apart from adversarial training, various technological methods exist to boost the robustness against adversarial attacks on machine learning models.

- **Defensive Distillation:** This method trains a smaller "student" model to copy the smoothed outputs (probabilities) of a bigger "teacher" model that has already been trained. The "softened" probabilities from the teacher model, which show how sure the teacher is about each class, are utilized as training labels for the student model. This procedure can make the student model less responsive to modest changes in input, which makes it difficult for an attacker to produce successful adversarial instances (Goldblum et al., 2020).
- **Input Transformation/Preprocessing:** Applying transformations to inputs before they reach the ML model to remove or mitigate adversarial noise.
  - Feature Squeezing involves reducing the color depth of images or spatial resolution to eliminate imperceptible noise (Wang et al., 2024). This technique operates on the principle that adversarial perturbations often reside in specific, subtle features that can be "squeezed out" without significantly altering the legitimate content of the image. For instance, it combines color bit depth reduction and spatial smoothing for detection. The method is designed to reduce the search space available to an adversary by coalescing samples that correspond to many different feature vectors into a single sample (Wang et al., 2024).

  - Randomized Smoothing involves adding random noise to inputs and then averaging the predictions of the model over multiple noisy versions to provide a probabilistic certificate of robustness.
  - JPEG Compression involves applying a non-differentiable compression algorithm which can often remove adversarial perturbations, as they are often highly sensitive to minute pixel values (Jia et al., 2020).
- **Gradient Masking:** Techniques that aim to make it difficult for an attacker to estimate the gradients of the model's decision function. While not a true defense against all attacks (as attackers might find non-gradient ways), it increases the challenge for gradient-based adversaries.
- **Certified Robustness:** An emerging area that aims to mathematically guarantee that a model's prediction will not change within a specified input perturbation bound. Unlike empirical defenses, these methods provide a provable guarantee. While currently not scalable for complex, large-scale deep learning models used in advanced humanoids, it's a promising research direction. It focuses on formally verifying that an AI model will behave as expected even when faced with subtle, malicious inputs. These methods are typically computationally intensive and currently limited to simpler models or specific types of perturbations, but they represent the gold standard of AI security. These methods will enable AI systems design for humanoids more resistant to malicious manipulation (Jia et al., 2020).

(4) **Real-Time Threat Detection and Monitoring**

Protecting a humanoid's safety from hostile attacks has to include real-time threat identification and monitoring (Hany et al., 2024).

- **Mechanism:** This involves deploying dedicated anomaly detection systems that continuously analyze a humanoid's sensor inputs, internal states, and action outputs. Any deviations from expected behavior – such as sudden changes in confidence scores from perception models, unusual motor commands, or inputs that differ statistically from normal operating conditions – can trigger alerts (Hany et al., 2024).
- **Application:** If developers can detect odd patterns in the robot's behavior or inputs, it might point to an attack. Should the robot start getting noticeably different from expected inputs, for example, the system can detect this as a possible attack and stop operations or change to a failsafe mode to protect the robot. This could involve activating a "safe mode" where the humanoid ceases autonomous operation, returns to a known safe state, or seeks human intervention. By use of proactive steps, humanoid developers may identify hostile activities in real time and react accordingly, therefore reducing the impact of these strikes.

(5) **Collaborative Defense Mechanisms**

In machine learning security, cooperative defensive techniques are quite crucial, considering the often changing and complicated character of hostile threats.

- **Federated Learning:** This approach facilitates the collaborative development of ML models without ever truly exchanging any raw data. Each humanoid (or a cluster of humanoids) trains a local model on its own data, then only sends encrypted or anonymized model updates (e.g., gradients or weights) to a central server. The server aggregates these updates to create a global model, which is then sent back to the humanoids. This helps mitigate privacy concerns and limits data exposure while still allowing for collective learning. Constant updating of the AI models of humanoids with fresh input from several sources via federated learning will enable the robot to be better protected against hostile attacks.

- **Threat Intelligence Sharing:** Partnerships with other businesses, research institutions, and AI security professionals are vital. This allows for the sharing of threat intelligence regarding new adversarial attack techniques and effective defense strategies, enabling the community to keep ahead of new risks and create more robust defense plans collectively.

Ethical and Societal Considerations

Because it makes sure that a humanoid robot is safe and reliable, machine learning security raises important moral and social issues. As humanoids are used more and more in businesses, public places, and even homes, security breaches are becoming more and more important. An aggressive attack that works on an industrial automation robot might cost a lot of money, cause a lot of downtime, or possibly put human workers nearby in great danger. Sensitive uses like healthcare and eldercare are much more dangerous since a hacked robot may put vulnerable people at risk by misreading vital signs or giving them the wrong care.

Questions also arise about responsibility when security failures occur in humanoids and similar AI-enabled systems. If an attempted attack on a humanoid is unsuccessful but still causes damage or harm, who should be held accountable? Is it the original manufacturer, the creator of the specific part (such as the AI model developer), the person who is putting the robot to work, or even the person who gave the training data? As more and more AI-powered robots enter society and hostile attacks become more complicated and harder to detect, these ethical and legal problems need to be dealt with ahead of time through explicit responsibility frameworks and strong industry standards. For the public to trust a humanoid, it will be important for them to be open about how their ML model works and what security precautions they have in place.

In conclusion, the integration of MLSec and the pursuit of adversarial robustness are not just technological goals for humanoid robots; they are foundational imperatives for their safe, reliable, and ethical deployment. As humanoids become more autonomous, ubiquitous, and integral to our lives, their ML systems will inevitably become targets. By proactively understanding the evolving threat landscape, investing in cutting-edge defensive strategies, and embedding security throughout the entire ML lifecycle, we can strive to ensure that humanoids fulfill their immense potential as trusted companions and invaluable tools, rather than becoming vectors for unforeseen harm or controversy. The commitment to responsible AI development, with ML security at its heart, will be the ultimate determinant of their successful integration into society.

7.4 Role of Regulation and Standards (ISO, IEEE, NIST)

Securing the complex humanoid ecosystem, as explored through best practices and the intricate AI safety frameworks, ultimately requires a robust foundation of universally accepted guidelines and, where necessary, enforceable regulations. Given the profound risks humanoids can pose, a patchwork approach to security is simply insufficient. The role of international standardization bodies and governmental agencies, such as the International Organization for Standardization (ISO), the Institute of Electrical and Electronics Engineers (IEEE), and the National Institute of Standards and Technology (NIST), becomes paramount in establishing the benchmarks for safety, security, and ethical operation. Their collective efforts aim to foster trust, ensure interoperability, and mitigate the dangers inherent in these intelligent, physically embodied machines.

The Imperative for Standardization and Regulation

The rapid advancement of robotics, particularly in the domain of humanoids, has created a regulatory lag. Traditional safety standards, often designed for static machinery or industrial robots in caged environments, are often inadequate for humanoids designed to operate autonomously in dynamic, human-centered spaces. The unique challenges posed by their physical presence, sophisticated AI, and potential for complex human interaction necessitate dedicated frameworks that address cybersecurity vulnerabilities, the integrity of AI decision-making, and the ethical implications of their autonomy. Without clear standards, manufacturers might adopt disparate security measures, leading to fragmentation, interoperability issues, and, ultimately, a higher risk of incidents that erode public confidence. Regulation, informed by these standards, provides the legal teeth, mandating compliance and establishing accountability, thereby pushing the entire industry towards a higher, safer baseline.

Key Players in Shaping Robot Security Standards

Several global organizations are at the forefront of developing the standards and frameworks critical for securing the humanoid ecosystem. Their work is often collaborative, drawing expertise from academia, industry, and government to build consensus on best practices.

ISO is a global leader in developing and publishing international standards across various industries, including robotics. While ISO's initial focus was heavily on industrial robot safety, particularly through the widely adopted ISO 10218 series (Robots and robotic devices – Safety requirements for industrial robots), their scope has evolved. ISO 10218, in its Parts 1 and 2, outlines detailed requirements for the design, manufacture, integration, and use of industrial robots, covering aspects such as safeguarding, emergency stops, and risk assessments. Crucially, recent revisions to ISO 10218, such as the 2023 updates, have begun to explicitly integrate cybersecurity requirements, recognizing that a physically safe robot is not truly safe if its control systems are vulnerable to digital attack. These updates often reference other security standards, such as the International Electrotechnical Commission (IEC) 62443 series for the security of industrial automation and control systems, thereby extending a comprehensive security framework to industrial robotics. Beyond industrial applications, ISO has also developed standards like ISO 13482 (Robots and robotic devices – Safety requirements for personal care robots), which addresses the unique safety considerations for robots designed to operate in close proximity to, or even physically interact with, humans in non-industrial settings, a highly relevant standard for consumer and service humanoids. These standards aim to ensure that robots meet critical safety benchmarks, protecting both humans and the broader systems they interact with, laying the groundwork for security through design and operation.

A particularly significant recent development from ISO, in collaboration with the IEC, is ISO/IEC 42001:2023, the world's first international standard for Artificial Intelligence Management Systems (AIMS). Published in December 2023, this standard provides a comprehensive framework for organizations to establish, implement, maintain, and continually improve an AI management system. For humanoid robots, which are inherently complex AI systems, ISO/IEC 42001 offers a structured approach to governing their entire AI lifecycle, from ethical design and development to responsible deployment and continuous operation. It mandates processes for identifying, assessing, and mitigating AI-specific risks, including not only security vulnerabilities and data compromise but also concerns such as algorithmic bias, lack of transparency, and ensuring accountability in AI decision-making. By adopting ISO/IEC 42001, organizations can demonstrate their commitment to developing and utilizing AI systems in a trustworthy, transparent, and secure manner, thereby building stakeholder confidence and aligning with emerging global AI regulations.

IEEE is a pivotal organization, renowned for its technical standards across electrical engineering, computer science, and related disciplines. For robotics and AI, IEEE plays a crucial role in developing forward-looking standards that address emerging technological capabilities and ethical considerations, often predating formal regulations. The IEEE's work extends beyond physical safety to encompass the trustworthiness and ethical behavior of autonomous systems, which are central to the reliability of humanoid systems. For example, the IEEE P7000™ series of standards addresses various ethical concerns in system design, including transparency, algorithmic bias, and human control of autonomous systems. While not always directly prescribing security controls, these ethical standards often imply or necessitate robust security measures to prevent malicious actors from exploiting ethical vulnerabilities or biases. The IEEE Robotics and Automation Society actively contributes to discussions and publications on robot security, often focusing on advanced topics like secure HRI, resilient control systems, and the validation of AI safety. Their reports frequently highlight unique risks associated with humanoid form factors, such as the potential for overtrust or the physical instability of bipedal systems, advocating for new testing methodologies and metrics that can be universally applied to evaluate and certify safe operation in dynamic environments. The IEEE's approach tends to be comprehensive, aiming for standards that facilitate interoperability, safety, and efficiency across the entire robot lifecycle and diverse applications.

NIST, a non-regulatory agency of the United States Department of Commerce, plays a critical role in developing cybersecurity frameworks and guidelines that are widely adopted globally. While NIST does not issue mandatory regulations (unless adopted explicitly by law), its publications are often considered authoritative best practices. The NIST Cybersecurity Framework (CSF), recently updated to CSF 2.0, provides a flexible, risk-based approach for organizations of all sizes and sectors to manage and reduce their cybersecurity risks. Although not robot-specific, its core functions – Identify, Protect, Detect, Respond, Recover, and Govern – are directly applicable to the cybersecurity of humanoid robots. Organizations deploying humanoids can utilize the CSF to evaluate their current cybersecurity posture, identify gaps, and implement controls across the robot's entire operational environment, encompassing its networked components and cloud services.

More specifically for AI, NIST has also developed the AI Risk Management Framework (AI RMF). This framework offers a structured approach for managing the risks associated with AI systems, with a primary focus on trustworthiness. It emphasizes principles such as explainability, fairness, privacy, and security in AI. For humanoids, the NIST AI RMF guides developers and deployers in identifying potential risks (e.g., biases in perception models, vulnerabilities in decision-making algorithms), assessing their impact, and implementing mitigation strategies to address these risks. It promotes continuous monitoring of AI outputs to ensure fairness, robustness, and adherence to security principles, serving as a voluntary guide for responsible AI development and deployment that aligns with global best practices.

Challenges and Future Directions

Despite the invaluable contributions of ISO, IEEE, NIST, and other bodies, the development and adoption of robust regulations and standards for humanoid robot security face significant challenges. The most prominent is the rapid pace of technological evolution. Robotics and AI are rapidly evolving fields; standards bodies, which operate by consensus, often struggle to keep pace with new capabilities, vulnerabilities, and applications, resulting in a regulatory lag. This makes it challenging to develop guidelines that are both comprehensive and future-proof, yet agile enough to remain relevant.

Another challenge lies in striking a balance between innovation and safety and security. Overly prescriptive or burdensome regulations could stifle technological advancement, whereas a lack of clear guidance could lead to unsafe deployments. The goal is to create frameworks that provide a sufficient baseline of security without imposing unnecessary constraints on research and development. Furthermore, global harmonization of standards is crucial. Humanoids are likely to be developed and deployed across international borders, necessitating consistent regulations to ensure interoperability and to prevent manufacturers from seeking out jurisdictions with weaker standards. Achieving this harmonization requires extensive international collaboration and consensus-building.

Finally, the complexity and interdisciplinary nature of humanoid security mean that standards must integrate knowledge from diverse fields: traditional cybersecurity, robotics engineering, AI safety, ethics, and psychology (for HRI). This demands a high level of expertise and cooperation among various stakeholders.

In conclusion, the role of regulation and standards, particularly those established by organizations such as ISO, IEEE, and NIST, is indispensable for building a secure and trustworthy humanoid ecosystem. These frameworks provide the essential benchmarks and guidelines for designing, developing, and deploying humanoids safely and securely. While challenges related to technological pace, balancing innovation, and global harmonization persist, their ongoing efforts are crucial in mitigating the profound risks identified, ultimately fostering public confidence and enabling the responsible integration of humanoids into society. Their work forms a critical external layer of defense and governance, complementing the internal best practices and architectural considerations for security.

Bibliography

Aleksieva-Petrova, A., & Petrov, M. (2023). Empirical study on secure software development. *2023 10th International Conference on Dependable Systems and Their Applications (DSA)*, 54–59. https://doi.org/10.1109/dsa59317.2023.00017

Bhatnagar, S., & Agrawal, R. (2024). Understanding explainable artificial intelligence techniques: A comparative analysis for practical application. *Bulletin of Electrical Engineering and Informatics*, *13*(6), 4451–4455. https://doi.org/10.11591/eei.v13i6.8378

Geppert, T., Deml, S., Sturzenegger, D., & Ebert, N. (2022). Trusted execution environments: Applications and organizational challenges. *Frontiers in Computer Science*, *4*. https://doi.org/10.3389/fcomp.2022.930741

Goldblum, M., Fowl, L., Feizi, S., & Goldstein, T. (2020). Adversarially robust distillation. *Proceedings of the AAAI Conference on Artificial Intelligence*, *34*(04), 3996–4003. https://doi.org/10.1609/aaai.v34i04.5816

Hany, Y., Kelg, Y., Emam, Y., El Bendary, N., Abdelshafy, Y., Alkady, G., & Mekhail, M. (2024). YOLOv8-Based surveillance robot for real-time threat assessment and mitigation. *2024 6th Novel Intelligent and Leading Emerging Sciences Conference (NILES)*, 577–580. https://doi.org/10.1109/niles63360.2024.10753171

Jia, J., Wang, B., Cao, X., & Gong, N. Z. (2020). Certified robustness of community detection against adversarial structural perturbation via randomized smoothing. *Proceedings of The Web Conference 2020*, 2718–2724. https://doi.org/10.1145/3366423.3380029

Macenski, S., Foote, T., Gerkey, B., Lalancette, C., & Woodall, W. (2022). Robot Operating System 2: Design, architecture, and uses in the wild. *Science Robotics*, *7*(66). https://doi.org/10.1126/scirobotics.abm6074

Park, S., Sim, J., & Park, J. (2019). System design of humanoid robot DYROS-JET. *2019 IEEE/SICE International Symposium on System Integration (SII)*, 746–750. https://doi.org/10.1109/sii.2019.8700453

Soundararajan, B. (2019). Proactive software development using secure by design principles. *International Journal for Multidisciplinary Research*, *1*(3). https://doi.org/10.36948/ijfmr.2019.v01i03.39048

Wang, Y., Li, X., Yang, L., Ma, J., & Li, H. (2024). ADDITION: Detecting adversarial examples with image-dependent noise reduction. *IEEE Transactions on Dependable and Secure Computing*, *21*(3), 1139–1154. https://doi.org/10.1109/tdsc.2023.3269012

Zafir, E. I., Akter, A., Islam, M. N., Hasib, S. A., Islam, T., Sarker, S. K., & Muyeen, S. M. (2024). Enhancing security of Internet of Robotic Things: A review of recent trends, practices, and recommendations with encryption and blockchain techniques. *Internet of Things*, *28*, 101357. https://doi.org/10.1016/j.iot.2024.101357

CHAPTER

8

Ethical Frontiers and Policy Considerations

Having thoroughly examined the cybersecurity threats and the essential frameworks for building secure and safe humanoid robots, we now pivot to the profound ethical frontiers and policy considerations that arise from their increasing autonomy and integration into human society. This chapter moves beyond the technical vulnerabilities of how a robot might be compromised to delve into the fundamental moral questions of how these intelligent machines should operate, interact with humans, and be governed, even when functioning as intended. The very capabilities that make humanoids so promising – their ability to perceive, learn, and make decisions in complex environments – also introduce a unique set of ethical challenges related to fairness, privacy, autonomy, and accountability. Addressing these issues proactively through thoughtful policy and design is crucial for fostering public trust and ensuring a beneficial future with humanoid robots.

8.1 Bias in AI and the Quest for Fairness

One of the most pressing ethical concerns with advanced AI systems, including those powering humanoids, is the potential for algorithmic bias. AI models learn from the data they are trained on. If this data reflects existing societal prejudices, historical inequalities, or unrepresentative demographics, the AI will inevitably perpetuate and even amplify those biases in its decisions and behaviors. For a humanoid robot, this could manifest in various ways.

DOI: 10.1201/9788743810704-11

- **Perceptual Bias:** A humanoid's vision system, trained predominantly on images of particular demographic groups, might struggle to accurately recognize or interpret the actions of individuals from underrepresented groups, leading to misidentification or delayed responses (Hundt et al., 2022). For example, a security humanoid might be less effective at detecting suspicious behavior in individuals whose appearance deviates from the biased instruction data on which it was trained.
- **Decision-Making Bias:** If a humanoid is used in applications such as hiring, loan approvals, or resource allocation (e.g., in a disaster scenario), and its decision-making AI was trained on biased historical data (Shuford, 2024), it could inadvertently discriminate against certain genders, ethnicities, or socioeconomic groups. For instance, a humanoid tasked with recommending candidates for a job might systematically favor applicants whose profiles align with historically privileged demographics, simply because its training data showed a correlation, not causation.
- **Interactional Bias:** A humanoid's language models or social interaction algorithms, if trained on biased conversational data, might exhibit subtle discriminatory language, perpetuate stereotypes, or respond differently based on a user's accent, gender, or perceived social status (Çırtlık & Cosar, 2024). This could lead to a less inclusive or even offensive user experience for some individuals.

The quest for fairness in AI involves a multi-faceted approach. It necessitates diverse and representative training data that accurately reflects the complexity of human populations, actively identifying and mitigating biases within datasets. Developing bias detection and mitigation techniques within AI algorithms themselves and implementing regular audits of deployed AI systems to monitor for discriminatory outcomes, are also crucial. Furthermore, establishing human oversight and review processes for high-stakes AI decisions can act as a critical safeguard against unintended bias.

Surveillance, Data Privacy, and the Right to be Forgotten

Humanoid robots are inherently sophisticated platforms for data collection. Equipped with an array of sensors – cameras, microphones, LiDAR, haptic sensors, biometric scanners – they continuously perceive and record their environment and the interactions within it. This pervasive data collection raises profound concerns about surveillance and data privacy.

- **Ubiquitous Monitoring:** A humanoid operating in a home, workplace, or public space could continuously record video, audio, and spatial data, creating a detailed digital footprint of individuals' lives, conversations, and movements. This raises questions about who has access to this data, how it's stored, and for what purposes it might be used.
- **Sensitive Data Collection:** Beyond general environmental data, humanoids may collect highly sensitive personal information, including biometric data (such as facial recognition

and gait analysis), health data (when used in care settings), or even emotional states inferred from vocal tones or facial expressions. The aggregation of such data could create detailed and potentially exploitable profiles of individuals.

- **Lack of Consent and Transparency:** In many scenarios, individuals interacting with a humanoid may not be fully aware of the extent of data being collected, how it's processed, or with whom it's shared. Obtaining meaningful consent for such pervasive data collection, especially in public spaces or for passive observers, is a significant challenge. Ensuring transparency about a robot's data practices is paramount.
- **The Right to be Forgotten:** As humanoids collect and store vast amounts of information, the concept of the "right to be forgotten," the ability for individuals to request the deletion of their data, becomes complex. How is this enforced across distributed robotic systems, cloud backups, and potentially federated learning models?

Addressing these privacy concerns requires a commitment to Privacy by Design, building data protection into the robot's architecture from the outset. This includes data minimization (collecting only essential data), anonymization or pseudonymization of sensitive information, robust encryption for data at rest and in transit, and strict access controls. Clear, easily understandable privacy policies and user-friendly consent mechanisms are crucial, enabling individuals to make informed decisions about their data. Regulatory frameworks, such as GDPR, provide a baseline, but specific adaptations are needed for autonomous, embodied systems.

The Challenge of Consent in Human-Robot Interaction

The concept of consent takes on new dimensions when applied to humanoid robots, particularly in social and personal contexts. The distinction between explicit and implicit consent becomes particularly complex in these interactions. While a user might explicitly consent to a robot performing a specific task, their implicit consent to continuous data collection or ongoing interaction might be far less clear. For instance, when someone invites a humanoid robot into their home, this action may seem like clear permission for the robot's presence, but it raises questions about whether this invitation also implies consent for the robot to map the living space, record conversations, or collect other forms of ambient data about the household.

The challenge becomes even more pronounced when considering vulnerable populations. Children, the elderly, and individuals with cognitive impairments face particular difficulties in providing truly informed consent for interaction with humanoid robots or for the data these systems collect. These groups may not fully comprehend the long-term implications of their interactions, the extent of data being gathered, or how that information might be used. This creates significant risks of manipulation or exploitation, as these populations may be more susceptible to the robot's designed appeal or may struggle to understand the complex technological and privacy implications of their engagement.

Third-party consent presents another layer of complexity that current legal and ethical frameworks struggle to address adequately. When humanoid robots operate in public or semi-public spaces such as hospital waiting rooms, retail stores, or community centers, they inevitably observe and potentially record individuals who have not explicitly agreed to such monitoring. These bystanders find themselves passively subject to data collection and behavioral analysis without having provided clear consent, creating a complex legal and ethical grey area that existing privacy laws may not adequately cover.

Perhaps most concerning is how the perceived autonomy and sophistication of humanoid robots can undermine the consent process itself. As these systems become more capable of seemingly intelligent or empathetic responses, humans naturally tend to anthropomorphize them, attributing intentions, understanding, or emotional capacity that the robots do not actually possess. This anthropomorphizing can lead people to overestimate the robot's capabilities or develop a false sense of trust and relationship with the machine. When individuals believe they are interacting with something approaching human-like consciousness or emotional understanding, their ability to provide truly informed consent may be compromised, as they may not fully grasp that they are engaging with a sophisticated but ultimately programmed system rather than a sentient being.

Policies must explore mechanisms for tiered consent, allowing users to grant different levels of permission for data collection and interaction. Clear visual and auditory cues from the robot can help indicate when it is recording or performing specific actions. Furthermore, robust opt-out mechanisms and the ability to easily review and delete collected data are crucial for empowering users.

In conclusion, the ethical frontiers of humanoid robots extend far beyond mere technical security. They compel us to confront fundamental questions about fairness, privacy, and consent in an increasingly automated world. Addressing algorithmic bias, safeguarding sensitive data, and ensuring meaningful consent are not just technical challenges but societal imperatives that demand careful consideration in policy, design, and public discourse. These ethical considerations form the bedrock upon which future laws and regulations for humanoids must be built, a topic we will delve into in the next section.

8.2 Existing Laws and Regulatory Gaps

Having explored the inherent ethical challenges posed by humanoid robots, particularly in relation to bias, surveillance, data privacy, and consent, it becomes imperative to examine the existing legal landscape and identify the significant gaps that must be addressed. Current legal frameworks, developed mainly before the widespread emergence of highly autonomous and intelligent robots, often struggle to adequately encompass their unique characteristics and the complex scenarios they present. While some existing laws offer partial coverage, the need for comprehensive, forward-looking regulation is increasingly evident, especially given the rapid pace of technological advancement.

Existing Legal Frameworks: Partial Fit, Imperfect Match

A range of existing legal frameworks can, to some extent, be applied to humanoid robots, typically by classifying them under existing categories like "products," "services," or "data processing entities."

Data Protection and Privacy Laws: In the EU, the GDPR is the cornerstone of data protection and privacy. It applies directly to humanoid robots that process personal data, which almost all do through their sensors and interactions with the environment. GDPR mandates principles such as data minimization, purpose limitation, transparency, lawful basis for processing, and data subject rights (e.g., access, rectification, erasure, and the right to object to automated decision-making). For humanoids, this means developers and deployers must consider the following.

- How do they obtain explicit consent for data collection, especially from individuals interacting with the robot?
- The legal basis for processing all personal data (e.g., legitimate interest, necessity for contract).
- Ensuring transparency about what data is collected, how it's used, and who it's shared with.
- Implementing robust security measures to protect this data.
- Facilitating data subject rights, which can be complex for data embedded in learning AI models.

GDPR wasn't designed for sentient, autonomous data collectors, leaving ambiguities, especially concerning passive data collection in public spaces or the complex "right to be forgotten" when data is integral to an AI's learning process.

Product Liability Law: In the UK, the Consumer Rights Act 2015 implements the EU's Product Liability Directive (though the UK's version remains in effect post-Brexit). This law holds manufacturers strictly liable for damage caused by a defective product. The challenge with humanoids is applying the definition of "defectiveness" to an AI system that learns and evolves over time. Is a learned, harmful behavior a "defect" if it wasn't present at the point of manufacture? What if the "defect" arises from a third-party software update or the interaction of multiple autonomous components? The traditional model struggles to attribute fault when an AI makes an unpredictable decision or when its behavior emerges from complex machine learning rather than a pre-programmed flaw. The EU's proposed AI Liability Directive aims to address these specific challenges by making it easier for claimants to prove fault and causation in cases of damages caused by AI systems. This is often achieved by introducing presumptions of defectiveness or causation in certain difficult-to-prove scenarios, as well as mandating the disclosure of evidence from AI providers.

Health and Safety Legislation: Laws such as the UK's Health and Safety at Work 1974 Act impose duties on employers to ensure the safety of their employees and others who may be affected by their work activities. This applies to humanoids in workplaces, mandating risk assessments and control measures. Similarly, general public safety laws would apply to humanoids operating in public spaces. However, these laws don't typically account for the unique risks of autonomous decision-making or the nuanced HRI that humanoids present.

Non-Discrimination Laws: Laws such as the UK's Equality Act 2010 prohibit discrimination based on protected characteristics (e.g., age, disability, race, gender). If a humanoid's AI exhibits bias that leads to discriminatory outcomes (as discussed in Section 8.1), these laws could be invoked. The challenge lies in proving intent or identifying the exact algorithmic cause of the discrimination within a complex AI system.

EU AI Act and the UK's Principles-Based Approach

Recognizing the limitations of existing laws, legislative bodies worldwide are developing dedicated frameworks for AI and robotics. The EU AI Act, which entered into force in August 2024, is the world's first comprehensive legal framework for AI. It adopts a risk-based approach, classifying AI systems into four levels: unacceptable, high, limited, and minimal risk (Schuett, 2023). Humanoid robots are highly likely to fall into the "high-risk" category, particularly if used as safety components in critical infrastructure, in employment, education, law enforcement, or as biometric identification systems. For "high-risk" AI systems, the Act imposes stringent obligations on providers and deployers before they can be placed on the market or put into service. These obligations include:

- Implementing robust risk management systems.
- Ensuring high-quality training, validation, and testing datasets to minimize bias and ensure accuracy.
- Maintaining detailed technical documentation and logging capabilities for traceability.
- Providing clear and adequate information to users.
- Implementing appropriate human oversight measures.
- Designing for high levels of accuracy, robustness, and cybersecurity.
- Establishing a quality management system.

The EU AI Act directly addresses concerns like bias, transparency, human oversight, and cybersecurity for AI systems. While it doesn't create specific "robot law," its definition of "AI system" and its risk classification profoundly impact humanoid manufacturers and users within the EU, and potentially globally due to its extraterritorial reach (affecting non-EU companies whose AI systems impact people within the EU).

In contrast to the EU's more prescriptive AI Act, the UK has adopted a principles-based, pro-innovation approach to AI regulation. Rather than introducing a single, overarching AI law, the UK government's strategy involves empowering existing regulators (e.g., ICO for data, HSE for safety) to interpret and apply their current powers to AI, guided by five cross-sectoral principles: safety, security and robustness; appropriate transparency and explainability; fairness; accountability; contestability/redress. The aim is to create a flexible framework that can adapt as the technology evolves, without stifling innovation through rigid rules. This approach emphasizes responsible innovation and aims to build public trust by embedding ethical considerations into regulatory functions. While this offers adaptability, critics suggest it might lead to regulatory fragmentation, inconsistencies, or a lack of clear legal certainty compared to a dedicated AI Act. However, for UK businesses supplying to the EU, compliance with the EU AI Act remains essential.

United States Innovation-First AI Policy

In the United States, the regulatory landscape for AI has historically been characterized by a sector-specific, voluntary approach that prioritizes innovation over prescriptive regulation. This framework has typically evolved through executive orders and agency-specific guidance rather than comprehensive federal legislation, reflecting America's traditional preference for market-driven solutions and technological leadership through competitive advantage rather than regulatory constraint (White House, 2025).

A significant shift in this approach materialized on 23 July 2025 with the release of "Winning the Race: America's AI Action Plan," a comprehensive federal strategy that outlines over 90 policy actions designed to cement American dominance in AI development and deployment. This ambitious plan represents a decisive pivot toward an aggressively pro-innovation stance, emphasizing deregulation and strategic public-private partnerships as the primary mechanisms for maintaining technological supremacy in the global AI race.

The 2025 AI Action Plan operates through three interconnected strategic pillars that collectively aim to transform the American AI ecosystem. The first pillar, focused on accelerating innovation, represents a fundamental reimagining of the federal government's role in AI development. Rather than imposing extensive regulatory oversight, this approach prioritizes the systematic removal of what policymakers characterize as bureaucratic obstacles and regulatory barriers that potentially slow private sector innovation. The strategy includes substantial increases in federal investment in AI research and development, the creation of streamlined pathways for bringing new AI solutions to market, and the establishment of robust public-private partnerships designed to develop what officials' term "secure, full stack AI export packages."

Particularly noteworthy is the plan's explicit emphasis on promoting open-source and open-weight AI models, reflecting a philosophical commitment to democratizing AI development while maintaining American technological leadership. Perhaps more controversially, the strategy includes directives to revise the globally recognized National Institute of Standards and Technology AI Risk Management Framework, specifically removing references to concepts such as misinformation mitigation, diversity and inclusion considerations, and climate change impacts. These revisions signal a narrower focus on technical performance and economic competitiveness over broader societal considerations.

The second pillar addresses the critical infrastructure requirements necessary to support large-scale AI development and deployment. This comprehensive approach to building American AI infrastructure encompasses the modernization and expansion of data center capacity, significant enhancements to the national power grid to support energy-intensive AI workloads, and improved access to high-performance computing resources for both private industry and academic research institutions. The plan includes specific directives to streamline federal permitting processes for

AI infrastructure projects, potentially reducing the time and regulatory complexity associated with building the massive computing facilities required for advanced AI development. Additionally, this pillar emphasizes ensuring access to high-quality, secure datasets necessary for training sophisticated AI models, recognizing that data availability often represents a critical bottleneck in AI advancement.

The third pillar focuses on leveraging AI capabilities to strengthen American leadership in international diplomacy and national security. This comprehensive approach aims to solidify United States' dominance in the global AI ecosystem by aggressively promoting the export of what officials term the "American AI Technology Stack," which encompasses not only hardware and software components but also technical standards, best practices, and governance frameworks. The strategy includes intensified enforcement of export controls designed to protect American AI innovations from potential adversaries while simultaneously facilitating technology transfer to allied nations. Furthermore, this pillar incorporates strengthened biosecurity protocols related to AI applications, recognizing the potential dual-use nature of advanced AI technologies in biological research and biotechnology applications.

The implementation of the AI Action Plan has been accompanied by several significant executive orders that provide concrete mechanisms for achieving these strategic objectives. These orders include specific directives for promoting the development and international export of American AI technologies, streamlining federal permitting processes for data center construction and expansion, and establishing new frameworks for public-private collaboration in AI development. The administration's comprehensive focus on deregulation and market-driven innovation reflects a fundamental belief that reducing regulatory friction will accelerate economic growth, technological advancement, and competitive positioning in the global marketplace.

However, this approach has generated considerable debate among policy experts, technology researchers, and civil society organizations. Critics argue that the emphasis on deregulation, particularly the specific revisions to established frameworks like the NIST AI Risk Management Framework, could potentially weaken important safeguards designed to address societal impacts such as algorithmic bias, privacy protection, and environmental sustainability. Some experts express concern that prioritizing speed and competitiveness over comprehensive risk assessment might lead to the deployment of AI systems that have not been adequately evaluated for their broader social implications.

Additionally, the federal government's deregulatory approach has contributed to increased regulatory fragmentation at the state level, as individual states continue to develop their own AI legislation to address concerns that federal policy may not adequately address. This creates a complex and often contradictory mosaic of AI governance across different jurisdictions, potentially complicating compliance for companies operating nationally while creating uncertainty about which standards and requirements will ultimately prevail.

Despite these concerns, proponents of the approach argue that maintaining American technological leadership in AI is essential for both economic prosperity and national security, and that excessive regulation could drive innovation and investment to more permissive jurisdictions. This ongoing tension between innovation promotion and risk mitigation continues to shape the evolving landscape of AI governance in the United States, creating a dynamic environment where federal deregulation coexists with state-level experimentation in AI oversight and control.

8.3 Critical Gaps and What is Missing?

Despite existing laws and emerging frameworks, several significant legal and policy gaps remain concerning humanoid robots, creating a complex landscape where technological advancement far outpaces regulatory development.

The question of legal personhood and agency represents one of the most philosophically and legally challenging frontiers in robotics law. As humanoid robots become increasingly sophisticated, capable of learning, adapting, and making complex decisions with minimal human oversight, legal systems worldwide grapple with whether these entities should ever be granted any form of legal recognition beyond their current classification as property. While most current legal frameworks firmly categorize robots as sophisticated tools or products, the emergence of truly autonomous behavior challenges traditional notions of agency and responsibility. The concept of "electronic personhood" has begun to surface in academic and policy discussions, though these conversations have primarily focused on practical considerations such as liability management and asset ownership rather than the more contentious possibility of granting human-like rights or moral status to artificial entities.

Perhaps the most pressing and complex gap lies in establishing accountability for autonomous actions. When a highly autonomous humanoid causes harm through its AI's unpredictable learning processes, an unforeseen interaction with its environment, or a decision made entirely without direct human input, the question of legal responsibility becomes extraordinarily difficult to resolve. The traditional chain of accountability becomes murky when dealing with systems that can modify their behavior, learn from experiences, and make decisions that their creators never explicitly programmed or anticipated. Current product liability laws struggle to address these scenarios because proving a "defect" in a continuously learning system or establishing a clear causal link between specific human actions and the robot's harmful behavior presents unprecedented challenges. While initiatives like the EU's proposed AI Liability Directive attempt to address some of these issues by reducing the burden of proof on victims, the fundamental philosophical and legal question of assigning moral and legal responsibility for truly autonomous behavior remains largely unresolved.

The development of standardization and certification frameworks for advanced autonomous systems reveals another critical gap in our regulatory infrastructure. Although organizations like ISO, IEEE, and NIST are actively developing standards for robotics and AI systems, comprehensive mandatory certification schemes specifically designed for high-risk autonomous humanoids remain immature and lack global harmonization. The challenge extends beyond simply creating standards to developing clear, auditable metrics and rigorous testing protocols that can reliably verify an AI system's adherence to safety, fairness, and transparency principles across the vast array of diverse real-world scenarios these robots will encounter. The complexity of certifying systems that learn and adapt continuously presents unique challenges that traditional product certification processes were never designed to handle.

The fundamental "pacing problem" between technological innovation and regulatory response creates an ongoing tension that undermines effective governance of humanoid robotics. Legislation, by its very nature, moves deliberately and cautiously, requiring extensive consultation, debate, and revision processes that can span years. In stark contrast, innovations in AI and robotics advance at an exponential pace, with breakthrough capabilities emerging in months rather than years. This temporal mismatch means that laws frequently become outdated before they're even fully implemented, as they attempt to regulate technologies and capabilities that didn't exist when the legislative process began. Some jurisdictions are experimenting with more agile regulatory mechanisms, such as "regulatory sandboxes" that allow controlled experimentation with emerging technologies under supervisory oversight, but these approaches remain limited in scope and uptake.

The absence of international harmonization in humanoid robotics regulation poses significant challenges for both innovation and governance. Given that these systems will inevitably operate across national borders and their components are sourced from global supply chains, the lack of international consensus on core regulatory principles creates a fragmented legal landscape. This fragmentation can lead to inconsistent safety and ethical standards, create barriers to innovation and trade, and potentially encourage "jurisdiction shopping" where companies seek out regions with the most permissive regulatory environments. The global nature of AI development and deployment makes coordinated international approaches essential, yet achieving such coordination remains politically and practically challenging. This fragmentation is further underscored by the diverse strategies adopted by major global powers. While the EU has pursued a comprehensive, risk-based legislative framework (as discussed in Section 8.2), and the UK a principles-based, delegated approach, the United States has recently formalized its own distinct strategy.

A significant development on 23 July 2025, saw the release of the Trump Administration's "Winning the Race: America's AI Action Plan." This plan outlines over 90 federal policy actions, marking a decisive pivot towards aggressively promoting U.S. dominance in AI primarily through deregulation and an industry-partnered environment. The plan is structured around three main pillars: Accelerating Innovation (focusing on removing perceived "red tape" to unleash private sector AI, increasing federal R&D, and promoting open-source AI models, alongside a directive to revise the NIST AI Risk Management Framework to remove references to specific societal concerns like "misinformation" and "Diversity, Equity, and Inclusion"), Building American AI Infrastructure (prioritizing data center capacity, grid modernization, and streamlined federal permitting), and Leading in International Diplomacy and Security (promoting the export of the "American AI Technology Stack" and intensifying export control enforcement). Accompanying executive orders further cement this focus on deregulation and market-driven growth. Critics, however, suggest that this approach, particularly its emphasis on deregulation and specific revisions to AI safety frameworks, could potentially lead to less robust safeguards for societal impacts like bias and environmental concerns, or contribute to regulatory fragmentation at the state level. This divergent policy philosophy from the US further complicates efforts towards global regulatory harmonization, creating a dynamic where varying national priorities contribute to the gaps and inconsistencies in AI governance.

The broader societal implications of humanoid robots, particularly their potential impact on employment and economic structures, reveal another significant gap in our legal and policy frameworks. As these systems become more capable of performing complex tasks traditionally reserved for human workers, they raise profound questions for labor law, social safety nets, and public policy regarding workforce transition and economic displacement. Current employment laws were designed for human workers and traditional automation, not for intelligent systems that might gradually assume roles across virtually every sector of the economy. The challenge extends beyond simple job displacement to questions of how society should adapt its economic structures, educational systems, and social support mechanisms to accommodate a future where human and AI collaborate or compete in the workforce.

In conclusion, while current legal frameworks provide some foundation for governing humanoid robots, they represent an imperfect and often inadequate response to the challenges these technologies present. New, dedicated legislation such as the EU AI Act represents significant progress in establishing clear obligations for high-risk AI systems, yet fundamental questions remain unresolved regarding legal personhood, accountability for autonomous actions, and the development of agile, globally harmonized standards. The presence of divergent national strategies, exemplified by the recent US AI Action Plan, further highlights the complexity of achieving unified global governance. The challenge for policymakers extends beyond crafting effective regulations to balancing the need for robust risk mitigation and ethical oversight against the imperative to preserve innovation and technological progress. This balance requires not just new laws but new approaches to governance that can adapt to rapidly evolving technologies while ensuring that the development and deployment of humanoid robots remains aligned with fundamental human values and societal wellbeing. The ongoing discourse in this area transcends mere rule-making to encompass fundamental questions about the kind of future we want to create with intelligent, embodied machines and how we can ensure that future serves humanity's best interests.

8.4 Responsibility and Liability in Case of Failure or Harm

The increasing autonomy and integration of humanoid robots into our daily lives, while promising transformative benefits, also introduce one of the most profound legal and ethical challenges of our technological era: establishing clear responsibility and liability in the event of failure or harm. When a sophisticated, intelligent machine operating with genuine autonomy causes damage or injury, the fundamental questions of blame and financial accountability become extraordinarily complex. Traditional legal frameworks, meticulously crafted for a world dominated by human actors and predictable mechanical failures, often prove inadequate when confronted with the emergent behaviors of advanced AI systems and the intricate, multi-layered supply chains that characterize modern robotics development and deployment.

The challenge extends beyond simple product liability into uncharted territory where the boundaries between human responsibility and machine autonomy blur in ways that existing jurisprudence has never had to address. Unlike conventional products that fail in predictable ways, humanoid robots represent a new category of entity that can learn, adapt, and make decisions that their creators never explicitly programmed or anticipated, creating liability scenarios that challenge our most fundamental assumptions about causation and responsibility.

The Problem of Attribution in Autonomous Systems

At its core, the liability challenge for humanoids stems from the extraordinary difficulty in attributing fault when an autonomous agent causes harm. The complexity of these systems creates a web of potential failure points that makes traditional cause-and-effect analysis near impossible. Unlike a faulty automobile where a defect can often be traced back to a specific manufacturing error or component failure, or a vehicle where the driver's negligence can be clearly established through evidence and witness testimony, a humanoid robot's harmful actions can arise from a bewildering array of sources that may interact in unpredictable ways.

These potential sources of failure include hardware malfunctions that may not manifest until specific environmental conditions are met, software bugs that emerge only through particular sequences of operations, unforeseen interactions between complex AI algorithms operating simultaneously, biased training data that leads to discriminatory or harmful decisions in real-world scenarios, inadequate testing protocols that failed to identify edge cases, or even perfectly executed actions that produce ethically undesirable outcomes while remaining entirely within the robot's programmed parameters. The interconnected nature of these systems means that a single incident may result from multiple contributing factors operating across different layers of the technology stack.

Consider a scenario in which a humanoid robot, tasked with assisting elderly patients in a busy hospital ward, collides with a frail patient while navigating a crowded corridor, resulting in serious injury. The immediate questions that arise illustrate the complexity of attribution in autonomous systems. Was the incident caused by a mechanical malfunction, such as a faulty brake system that failed to engage when the robot detected an obstacle, a compromised sensor that provided inaccurate distance measurements, or a degraded actuator that prevented precise movement control? Such hardware failures might point clearly to the component manufacturer or the system integrator responsible for quality assurance.

Alternatively, the collision might have resulted from a software error, such as a bug in the robot's navigation algorithm that miscalculated the optimal path through the corridor, a flaw in its object detection system that failed to properly classify the patient as a human requiring special consideration, or a failure in its safety protocols that allowed it to operate at inappropriate speeds in a crowded environment. These software-related causes could implicate the original software developer, the team responsible for system integration, or even the entity that customized the software for the specific hospital environment.

More troublingly, the incident might have resulted from an AI decision-making failure, where the robot's machine learning systems made what could be considered an "unforeseen decision" based on its training and operational parameters. Perhaps the AI prioritized completing a time-sensitive medication delivery over patient safety in a novel situation that wasn't explicitly addressed in its programming, or it made an autonomous judgment about acceptable risk levels that proved tragically incorrect. Such scenarios raise profound questions about the responsibility of the AI model developer, the entity that provided the training data, the organization that defined the robot's operational priorities, and even the broader AI research community that developed the underlying algorithms.

The situation becomes even more complex when considering whether the incident resulted from inadequate deployment or maintenance practices. Was the robot operating in an environment or under conditions that exceeded its design specifications? Did the hospital staff fail to apply critical software updates that would have prevented the collision? Were proper safety protocols and training procedures followed by human operators? Such questions might shift liability away from the manufacturers toward the robot's owner, operator, or the institution responsible for its deployment and maintenance.

Finally, the possibility of malicious compromise adds another layer of complexity to liability attribution. If the robot was compromised by a cyberattack that altered its behavior or disabled its safety systems, the immediate responsibility might lie with the attacker. However, this scenario raises additional questions about the cybersecurity diligence of both the manufacturer and the operating institution. Did they implement adequate security measures, follow best practices for system updates and monitoring, and respond appropriately to known vulnerabilities?

The traditional legal frameworks of strict liability and fault-based negligence, while sophisticated in their own domains, struggle to address these multifaceted attribution challenges. Strict liability, which holds manufacturers liable for product defects regardless of fault, works effectively when dealing with tangible, identifiable defects in conventional products. However, defining a "defect" in an AI system that continuously learns and evolves after leaving the factory presents unprecedented conceptual and practical difficulties. How do courts determine whether an AI's decision represents a defect in its programming or training, or simply an unfortunate but reasonable response to an unforeseen situation?

Similarly, proving negligence becomes extraordinarily challenging when dealing with AI systems whose decision-making processes are often opaque or "black box" in nature. Traditional negligence requires demonstrating that a party's actions or inactions fell below a reasonable standard of care, but establishing what constitutes reasonable care for AI development, deployment, and maintenance remains largely undefined. When an AI's behavior emerges from complex interactions between millions of parameters trained on vast datasets, determining whether any human actor failed to meet an appropriate standard of care becomes a formidable legal and technical challenge that existing judicial processes are ill-equipped to handle.

Assigning Responsibility Across the Value Chain

The development and deployment of a humanoid robot typically involve a long and complex value chain, making the assignment of responsibility even more intricate. Each actor within this chain bears a degree of potential responsibility.

- **Manufacturers and Hardware Developers:** Primarily responsible for physical defects, design flaws in the robot's hardware, and ensuring the robot's mechanical safety systems are robust and tamper-proof. Their liability would typically fall under product liability laws.
- **Software and AI Developers:** Responsible for the code that runs the robot and the algorithms that govern its intelligence. This includes ensuring the software is free of bugs, the AI model is robust against unforeseen inputs, and the training data is free from biases that could lead to harmful outcomes. Their liability can be harder to establish, especially if the harm results from an emergent AI behavior rather than a direct coding error.
- **Integrators and Deployers:** Often, third-party companies integrate a humanoid into a specific environment (e.g., factory floor, hospital) and customize its operation. They are responsible for ensuring the robot is installed correctly, that its operating environment is suitable, and that it is adequately tested for its specific tasks. Their negligence in integration or testing could lead to liability.

- **Owners and Operators:** These are the end-users of the humanoid. Their responsibilities typically include ensuring the robot is operated within its intended parameters, receiving adequate training, performing necessary maintenance, applying software updates, and providing appropriate supervision. Negligent operation or a failure to maintain the robot could lead to their liability.
- **Third-Party Actors (Hackers):** In cases of cyberattack, the malicious actor causing the compromise would bear primary criminal and potentially civil liability. However, this does not necessarily absolve other parties (manufacturer, owner) if their cybersecurity defenses were negligent or failed to meet industry standards.

The challenge lies in the "shared control" paradigm, where human operators might supervise or even co-pilot the robot, blurring the lines of who is in control at any given moment. Furthermore, the inherent opacity of complex AI models (the "black box" problem) makes it incredibly difficult to trace the causal link between a specific input or internal state of the AI and a harmful output, complicating fault assessment.

Evolving Legal Approaches and Future Directions

Recognizing these gaps, legal systems are beginning to adapt, albeit slowly. The EU has been at the forefront of proposing new frameworks to address AI liability. The EU's proposed AI Liability Directive, for instance, aims to modernize liability rules for AI systems. It seeks to make it easier for victims to claim damages by addressing some of the challenges of proving fault or causation. Key proposals include:

- **Presumption of Causation:** In specific "high-risk" AI contexts (as defined by the EU AI Act), if a victim can prove that a relevant fault (e.g., non-compliance with a safety standard) occurred and that damage occurred, a causal link between the AI's output and the damage might be presumed, shifting the burden of proof partly to the AI provider.
- **Evidence Disclosure:** The Directive proposes mechanisms to allow victims to request access to evidence from AI providers, helping to overcome the "black box" challenge.

Beyond these specific directives, broader discussions include:

- **Mandatory Insurance Schemes:** Similar to car insurance, a system where mandatory insurance covers damages caused by robots, potentially without proving fault, could ensure victims are compensated quickly. This approach spreads the risk across the industry or society.
- **Risk-Based Liability:** Aligning liability with the risk classification of the robot (as seen in the EU AI Act), meaning higher-risk humanoids might face stricter liability rules.
- **"No-Fault" Funds:** Creating dedicated compensation funds that victims can claim from, funded by industry contributions or public money, without the need for lengthy legal battles to prove fault.

- **Certification and Auditing:** Strengthening certification processes and requiring regular audits of AI systems to ensure they meet safety and ethical standards, with non-compliance potentially leading to easier assignment of liability.

In the UK, while the government prefers a principles-based approach to AI regulation, the issue of liability remains a critical area of review. Existing product liability and negligence laws will remain in effect. Still, there is an ongoing recognition that these may need supplementation or clearer guidance to address the unique challenges of AI autonomy. The UK's emphasis on responsible innovation aims to encourage developers to consider liability from the design stage.

Ultimately, navigating the ethical and legal frontiers of responsibility and liability for humanoid robots requires a careful balance. Overly strict or unclear liability regimes could stifle innovation, while a lack of accountability could undermine public trust and leave victims without recourse. The future legal landscape will likely involve a blend of updated product liability, specific AI liability directives, and potentially new insurance models, all aimed at ensuring that as humanoids become more capable, the framework for managing their risks and ensuring justice in cases of harm evolves in tandem. This ongoing policy development is essential for a truly responsible integration of humanoid robots into society.

Bibliography

Çırtlık, B., & Cosar, S. (2024). Gender bias in AI. *Feminist Asylum: A Journal of Critical Interventions, 2.* https://doi.org/10.5195/faci.2024.124

Hundt, A., Agnew, W., Zeng, V., Kacianka, S., & Gombolay, M. (2022). Robots enact malignant stereotypes. *2022 ACM Conference on Fairness Accountability and Transparency*, 743–756. https://doi.org/10.1145/3531146.3533138

Schuett, J. (2023). Risk management in the Artificial Intelligence Act. *European Journal of Risk Regulation, 15*(2), 367–385. https://doi.org/10.1017/err.2023.1

Shuford, J. (2024). Exploring ethical dimensions in AI: Navigating bias and fairness in the field. *Journal of Artificial Intelligence General Science (JAIGS), 3*(1), 103–124. https://doi.org/10.60087/jaigs.vol03.issue01.p124

White House. (2025, July 23). White House unveils America's AI action plan – The White House. *The White House.* https://www.whitehouse.gov/articles/2025/07/white-house-unveils-americas-ai-action-plan/

CHAPTER 9

The Road Ahead

9.1 Predictions for Humanoid Integration in Society

Having explored the intricate layers of humanoid capabilities, their ethical challenges, and the essential frameworks for their security and governance, we now turn our gaze to the future. The question is no longer if humanoids will integrate into society, but how rapidly, in what capacities, and with what profound effects. Based on current technological trajectories, advancements in AI, and the burgeoning investment in robotics, we can anticipate a phased but increasingly pervasive integration of humanoids across various facets of human life. This will be a transformation far more nuanced and widespread than the industrial automation of the past.

Initially, humanoids will likely find their most impactful roles in environments where precision, endurance, and safety in hazardous conditions are paramount, or where repetitive tasks benefit from automation. We'll see an acceleration of their presence in manufacturing and logistics, moving beyond fixed industrial arms to mobile, adaptable humanoids capable of navigating dynamic factory floors, handling delicate components, and performing intricate assembly tasks alongside human workers. In hazardous environments, such as disaster zones, nuclear facilities, deep-sea exploration, or even space missions, humanoids will serve as invaluable proxies, mitigating risks to human life by undertaking dangerous inspections, repairs, and data collection. Their dexterity, combined with remote teleoperation capabilities, will make them indispensable tools where human presence is too risky or impossible.

DOI: 10.1201/9788743810704-12

Beyond these specialized industrial roles, the most significant shift will be the integration of humanoids into service industries and personal care. In healthcare, expect humanoids to evolve into sophisticated assistants, aiding nurses with patient monitoring, delivering supplies, and supporting surgical teams with precision tasks. For elderly care and individuals with disabilities, humanoids could offer unprecedented levels of assistance, from reminding about medication and helping with mobility to providing companionship and facilitating communication, thereby alleviating pressure on human caregivers and enhancing quality of life. In sectors such as retail and hospitality, humanoids could assume roles as customer service agents, concierges, or dynamic cleaners, offering personalized interactions and maintaining public spaces. Their ability to perceive human emotion and respond with nuanced social cues, while still in its nascent stages, promises a new era of interactive service.

In our homes and personal lives, the integration will be gradual but transformative. Early adoption might lead to humanoids being perceived as versatile domestic helpers, capable of performing a wider range of household chores than the current robotic vacuums and adapting to the unique layout of each home. Their ability to learn routines and anticipate needs could make them seamless members of the household ecosystem. More profoundly, as their social AI advances, they may fulfill roles as companions, offering interaction, stimulation, and a sense of presence, particularly for those experiencing loneliness or social isolation. However, this is an area where ethical considerations, such as authenticity, emotional connection, and potential over-reliance, will remain at the forefront.

Finally, in public spaces and urban infrastructure, humanoids could become ubiquitous. Imagine intelligent robots assisting with navigation in airports, providing real-time information in tourist hotspots, or performing intricate maintenance tasks on complex infrastructure, such as bridges and public transportation systems. They can also contribute to public safety and security by patrolling areas, monitoring for anomalies, and providing rapid responses in emergencies, acting as a force multiplier for human security personnel. As urban environments become increasingly intelligent, humanoids could serve as mobile nodes in smart city networks, collecting environmental data, managing traffic flows, and interacting with citizens to enhance urban living.

The integration of humanoids will also inevitably trigger significant societal shifts. Economically, it will reshape labor markets, necessitating new forms of education and training to prepare human workers for collaboration with and supervision of robotic colleagues. Socially, it will challenge our perceptions of intelligence, companionship, and even what it means to be human as we increasingly interact with machines that mimic human form and behavior. Urban planning may adapt to accommodate their movement, charging infrastructure, and public interaction points. While the precise timeline for these transformations remains uncertain and subject to technological advances, regulatory frameworks, and societal acceptance, the trajectory suggests humanoids will move from niche industrial tools to integrated, interactive members of our future society.

Humanoid robots will likely not dominate our lives but rather weave themselves into them. They will become another layer of the built environment, as unremarkable as elevators or streetlights are to us today. But this shift requires careful steps: solid design, clear ethical boundaries, and respect for human dignity. If we get those aspects right, the presence of humanoids in daily life will not feel like a takeover. It will feel like the natural next step in our long history of building tools that extend our abilities, the difference being that these tools will stand beside us, look us in the eye, and, in their way, understand what we need.

9.2 Will Humanoids Become Trusted Companions or Controversial Tools?

Listen into any conversation about humanoid robots and you'll hear two very different stories. In one, these machines are loyal companions, patiently helping us through daily life, quietly doing the tasks we don't enjoy, and offering comfort to the lonely. On the other hand, they are a looming threat – watching, recording, taking jobs, and blurring the line between human and machine in ways that feel unsettling. Both visions carry some truth. The path we choose will depend on how we design, regulate, and live with these machines in the years ahead.

The path to humanoids becoming trusted companions relies heavily on their demonstrated reliability and safety. As discussed in Chapter 7, robust cybersecurity measures and rigorous AI safety frameworks are non-negotiable. If early deployments are plagued by security breaches, unpredictable malfunctions, or unintended harm, public confidence will shatter, potentially setting back widespread adoption by decades. Trust is also fostered through transparency and explainability in their AI systems. If users can understand, at least at a high level, why a humanoid made a particular decision, it reduces the "black box" mystery. It builds a sense of predictability and control. This extends to their ethical behavior: humanoids that consistently demonstrate fairness, respect privacy, and operate within clear consent boundaries (as explored in Section 8.1) will be far more readily accepted than those perceived as invasive or biased. Effective HRI, where humanoids communicate, adapt to individual needs, and respond with appropriate social cues, will also play a crucial role in making them feel less alien and more approachable. Ultimately, trust will grow from demonstrated benefits and a clear value proposition, showing how humanoids genuinely enhance lives, solve real problems, and contribute positively to society, whether in caregiving, hazardous work, or daily assistance.

Conversely, several factors could lead humanoids to become controversial tools. The most immediate threat is the specter of security risks and hacks, echoing the concerns from Chapter 6. A widely publicized incident involving a compromised humanoid that causes significant harm or invades privacy could ignite widespread public outcry and resistance. Similarly, recurring ethical failures, such as persistent algorithmic bias leading to discrimination, blatant privacy violations, or perceived manipulation (as discussed in Section 8.1), would severely damage their public image. The looming concern of job displacement represents another significant source of potential controversy. If humanoids are perceived primarily as replacements for human labor, leading to widespread economic disruption without adequate social safety nets or new opportunities, societal backlash could be severe.

Furthermore, the ongoing challenge of a lack of accountability and clear liability (Section 8.4) will fuel controversy. If victims of robot-induced harm struggle to find legal recourse or hold responsible parties accountable, it will breed distrust and a sense of injustice. The inherent psychological discomfort of the "uncanny valley" effect, where humanoids resemble humans almost, but not entirely, can also contribute to unease and resistance. Beyond that, deep-seated fears about the loss of human control over increasingly autonomous systems (touched upon in Sections 6.3 and 7.2) persist, particularly amplified by science fiction narratives. The potential for weaponization or military use of humanoids also remains a significant ethical minefield, threatening to cast a shadow over all civilian applications.

Early experiences and incidents will largely determine the tipping point between trust and controversy. A series of positive, beneficial, and secure deployments could accelerate acceptance, while a few high-profile negative incidents could severely impede it. The effectiveness of regulatory bodies and standards (Sections 7.3 and 8.2) will also be critical; clear, robust, and adaptable frameworks can provide the necessary guardrails for safe development and deployment, mitigating risks before they become societal flashpoints.

Finally, the media portrayal and public education surrounding humanoids will significantly shape public opinion, either foster understanding and realistic expectations or fuel sensationalism and fear. Ultimately, the burden of ensuring trust lies squarely on the shoulders of developers, policymakers, and deployers, who must prioritize responsible innovation at every turn. Their collective commitment to ethical design, robust security, transparency, and clear societal benefits will ultimately determine whether humanoids are embraced as companions or relegated to the realm of controversial tools.

9.3 Final Thoughts on Balancing Innovation with Responsibility

As we stand on the precipice of a new era, one increasingly shaped by the presence and capabilities of humanoid robots, the overarching challenge is clear: how do we effectively balance the relentless drive for innovation with an unyielding commitment to responsibility? Throughout this book, we've journeyed from the historical origins of humanoids to their complex inner workings, explored the very real threats they face, and grappled with the profound ethical and legal questions their existence poses. The insights gleaned from these discussions converge on a singular truth: the future of humanoids and their ultimate impact on society hinges on our collective ability to navigate this delicate equilibrium.

Innovation is the lifeblood of progress. It pushes the boundaries of what's possible, promising humanoids that can revolutionize industries, provide unprecedented care, perform tasks beyond human capability, and even offer companionship to those in need. The breakthroughs in AI, robotics, materials science, and HRI are undeniable, opening doors to a future that once belonged solely to science fiction. To stifle this innovation through excessive caution or premature, rigid regulation would be to forgo immense potential benefits – a decision that itself carries a significant responsibility. The drive to create more capable, autonomous, and integrated humanoids will undoubtedly continue, propelled by economic opportunities, scientific curiosity, and the desire to solve complex human problems.

However, unchecked innovation, lacking a strong ethical compass and robust safeguards, carries inherent risks. We've seen how vulnerabilities can be exploited, leading to physical harm, privacy breaches, and economic disruption. We've grappled with the subtle yet pervasive dangers of algorithmic bias, the complexities of data privacy in an always-on world, and the thorny question of accountability when an autonomous machine makes an error. The potential for humanoids to exacerbate existing societal inequalities, displace workers without adequate transition, or be deployed in ways that erode human autonomy or dignity is real. Responsibility, in this context, means anticipating these challenges, designing with foresight, and building safeguards into the very core of these machines and the ecosystems in which they operate. Striking this balance demands a continuous, iterative dialogue and collaborative action from all stakeholders.

- Developers and Engineers bear the primary responsibility for Security and Safety by Design. This means not viewing cybersecurity and AI safety as optional add-ons, but as foundational pillars integrated from conception to deployment. It involves rigorous testing, transparent development practices, and a deep understanding of potential failure modes and ethical implications.
- Policymakers and Regulators must develop Agile and Adaptive Frameworks. They face the challenging task of creating regulations that are both flexible enough to keep pace with rapid technological change and robust enough to enforce accountability and protect fundamental rights. This requires a nuanced understanding of the technology, a willingness to engage in "regulatory sandboxes" for experimentation, and a commitment to international harmonization to avoid fragmented legal landscapes.
- Industry Leaders and Organizations must champion Ethical Deployment and Stewardship. This goes beyond mere compliance with laws, embracing a corporate culture that prioritizes responsible innovation, invests in ethical AI practices, ensures fair labor practices, and maintains open communication with the public about the capabilities and limitations of humanoid technology.
- The Public and Civil Society play a crucial role in Informed Discourse and Oversight. Public understanding and trust are paramount. Open dialogue, critical assessment of media narratives, and active participation in shaping policy debates are essential to ensure that the development and deployment of humanoids reflect societal values and aspirations, not just technological prowess.

Ultimately, the future of humanoids is not a predetermined destination but a path we forge together. It is a path that calls for courage to innovate, wisdom to anticipate risks, and an unwavering commitment to prioritizing human well-being and ethical principles. By consciously and continuously balancing the immense potential of humanoid innovation with a profound sense of shared responsibility, we can shape a future where these intelligent machines truly serve humanity, becoming trusted companions and powerful tools for positive change. The conversation has begun; the critical work of building that future, responsibly, now lies ahead.

Index

For Product Safety Concerns and Information please contact our EU representative GPSR@taylorandfrancis.com Taylor & Francis Verlag GmbH, Kaufingerstraße 24, 80331 München, Germany

Batch number: 10817815

Printed by Printforce, the Netherlands